SHELTER FROM THE STORM

SHELTER FROM THE STORM

HOW CLIMATE CHANGE IS CREATING
A NEW ERA OF MIGRATION

Julian Hattem

THE
NEW
PRESS

NEW YORK
LONDON

Requests for permission to reproduce selections from this book should be made
through our website: https://thenewpress.org/contact-us.

Published in the United States by The New Press, New York, 2026
Distributed by Two Rivers Distribution

ISBN 978-1-62097-847-4 (hc)
ISBN 978-1-62097-943-3 (ebook)
CIP data is available

The New Press publishes books that promote and enrich public discussion and
understanding of the issues vital to our democracy and to a more equitable world.
These books are made possible by the enthusiasm of our readers; the support of a
committed group of donors, large and small; the collaboration of our many partners
in the independent media and the not-for-profit sector; booksellers, who often hand-
sell New Press books; librarians; and above all by our authors.

www.thenewpress.org

Printed in the United States of America

10   9   8   7   6   5   4   3   2   1

For Max

Contents

SHELTER FROM THE STORM

Introduction

Where We Are Going

The water kept rising. For days, the rains in southern Brazil had been unrelenting, pounding down on fields and houses and filling the streams and rivers that crisscross the region. Heavy rains were usual this time of year, but the downpour refused to relent and offer the ground even a temporary reprieve. Rivers burst their banks, seeping first onto the shorelines and then into towns and villages. Dams filled to the breaking point. Water raced down hillsides and through streets, dragging along mud and stones and still continuing to gather force, triggering landslides that lurched onto roads and slammed into the sides of houses. Soon, the water swallowed cars, parts of buildings, and anything else left vulnerable to the elements. Bridges and roads were wiped aside.

And still it poured. In towns across Brazil's southernmost state of Rio Grande do Sul, eight months' worth of rain fell in two weeks.[1] As the rivers swelled, water raced downstream toward Porto Alegre, the prosperous state capital home to more than 4 million people and one of South America's most important industrial centers. The wealth of the city, with its grand architecture and large marinas, rests in part on its location where four rivers join the oversized, lakelike Guaíba River at the edge of a coastal lagoon. Surrounded by water, Porto Alegre was inundated from all sides. As the municipal defenses failed, residents awoke to find water filling the streets up to their ankles; within a few hours, it would be up to their waists, and then their roofs.[2] Before long, city streets turned into canals of putrid brown water, as emergency responders conducted rescues on motorboats and jet skis. Residents used inflatable mattresses to ferry their neighbors to safety. Entire neighborhoods were submerged, trash and treasured heirlooms alike scattered like so much flotsam. The water level of the Guaíba rose seventeen feet, far surpassing the record set in a historic 1941 flood.[3] Cars and trucks bobbed haphazardly like bath toys swept aside by an impatient toddler. The water filled the streets for two weeks.

In all, 90 percent of the state of Rio Grande do Sul was affected by the extreme rains. Nearly 1,500 square miles were flooded, an area larger than Rhode Island, sandwiched between Argentina, Uruguay, and the South Atlantic.[4] Hundreds of cities declared states of emergency. Around 775,000 people were displaced and 183 died.[5] The damages were calculated to be greater than $15 billion, with nine out of every ten businesses suffering partial or total losses.[6] For onlookers in the United States, the scenes evoked memories of New Orleans after Hurricane Katrina two decades earlier, when our own major southern city was inundated by water and irreparably marked.

As New Orleans residents did during Katrina, those fleeing Rio Grande do Sul gathered wherever they could. Hundreds crowded next to one another in gyms and schools and evacuation centers. They slept on the couches and floors of family members and friends. Some drove their cars up to higher ground and stayed there until it was safe to go somewhere else. Others erected tents in dry parking lots, found places underneath bridges, or camped out anywhere else that seemed safe. As the water rose, many who could not or did not evacuate

tried to outrun it by climbing to higher floors and were eventually rescued from their rooftops by helicopter.

For many residents of rural areas and poorer suburbs, the
flood was the latest in a string of tragedies. Other powerful
floods had struck just a few months earlier, and some people
had fled their homes three times in less than a year. Some insisted that this evacuation would be their last. "I have no idea
where I'm going, but it will be somewhere far from the river,
where our lives will not be at risk," one businessman told
journalists while dumping wheelbarrows full of mud from his
home in the small village of Muçum, ninety miles upriver from
Porto Alegre.[7] In better times, his village is surrounded by lush
greenery and farmland, as steep hills arch down to the Taquari
River. But the floodwater transformed much of the landscape
into a muddy red-brown smudge, erasing whatever roads and
bridges and houses lay in the way. After the repeated disasters,
residents decided to move on. "Many of our neighbors have
not even come back to clean up their homes and are looking
for other places to live," said another resident. "Honestly, we
have no tears left to cry."[8] The mayor of Muçum and several
other local officials talked openly about relocating entire neighborhoods, moving people first to temporary shelters and then
eventually to brand-new buildings farther from harm's way.[9]

Rio Grande do Sul's once-in-a-century flood was caused
by El Niño–related rains and linked to the region's location
at the meeting point of tropical and polar air. It was made
more than twice as likely by climate change.[10] Careless policy
was also to blame; the risk of disaster had been heightened
by shortsighted agricultural and deforestation practices and
poor urban planning that led to houses being constructed at
Porto Alegre's periphery, in natural areas that had previously

acted as a buffer against flooding and landslides. The city had also long neglected its five-decade-old anti-flood system, which failed when it was needed most. Instead, the network of dikes ended up preventing floodwater from draining into the Guaíba, effectively trapping it in the city and prolonging the crisis.[11] City officials said the network was never designed to combat a deluge of this size.[12] All the same, researchers had for years warned that rainfall was increasing and that Brazil's southern cities were not prepared for the possibility of disasters of this magnitude.[13] The defenses were relics of another climate era, infrastructure for a world that no longer exists.

The devastation of southern Brazil was an extraordinary moment for Latin America's most populous country, forcing national leaders to reckon with their failure to prepare for a climate that has grown increasingly hostile. But these kinds of disasters have come to be expected. Just a few days after Brazil was being overwhelmed, Cyclone Remal slammed into Bangladesh and India, dumping water, lashing buildings, and uprooting trees with winds that measured over seventy miles per hour. The storm landed in the Sundarbans delta, tearing down embankments that dot the coastal areas, washing out beachside cafés and fishing villages and farms alike.[14] Thatched roofs were blown off houses and trees scattered across power lines and roads. Dozens of people were killed. The deluge submerged roadways one hundred miles inland in Dhaka, Bangladesh's crowded capital, and left millions of people without power.[15] Across Bangladesh and India, about a million people were forced to flee to sturdy concrete storm shelters or other safety points. Many of them came from small villages where flimsy houses were totally destroyed by the storm.[16]

Meanwhile, on the other side of the globe, the Democratic Republic of the Congo was just beginning to recover from its worst flooding in six decades, as heavy rains swelled the Congo River above its banks. Half a million people fled when the water raced into their homes, and some people were forced to spend more than a month living in makeshift camps.[17] Intense rains would pelt countries across West and Central Africa for months, causing devastating flooding that displaced 4 million people over the course of the year.[18] Fleeing for their lives, people slept on highways and in gas stations—any place offering a small dry spot of ground—while they watched dead people and animals float by.[19] In Congo, Nigeria, Mali, and several other countries, the flooding followed years of crisis and conflict, as a long run of environmental changes, militant violence, and foreign exploitation has exacerbated people's poverty and insecurity.

Similar scenes unfolded across Europe and North America, too. After forming over northern Italy, Storm Boris cut through Central Europe, dropping as much as a foot and a half of rain in three days. In parts of Austria, the amount of rain nearly doubled the prior record.[20] Soon after, a different weather system crashed into eastern Spain, triggering flash floods that decimated the outskirts of Valencia in the country's worst natural disaster in a generation. A year's worth of rain fell in just a few hours, creating new urban rivers where streets had been.[21] Mud and water smashed cars and slammed into buildings and rail lines. In the United States, Hurricane Helene roared into Florida and left a trail of destruction up through North Carolina and Virginia; some of the hardest-hit areas were 350 miles inland, in mountainous communities in and around Asheville. Rain had drenched the soil around Asheville in the days before the storm system arrived, so rivers and streams were quick to overflow and send water racing through neighborhoods. More than 230 people were killed, in the mainland United States' deadliest hurricane since Katrina.[22] Just days later, Hurricane Milton touched down on some of the same parts of Florida, forcing millions to evacuate.

All these events occurred within a couple of months in 2024, amid a year of disasters that are too numerous to list comprehensively, as part of our planet's single warmest year on record—at least for now.[23] Globally, temperatures that year were 1.47°C (2.65°F) higher than the average during the pre-industrial times of the mid-nineteenth-century (from 1850 to 1900).[24] One Sunday that July was the single hottest day in recorded history, with global average air temperatures of 17.09°C (62.76°F).[25] That record was broken just a day later, when the average temperature reached 17.15°C (62.87°F); evidence suggests that Monday may have been the planet's hottest day in more than 100,000 years.[26] By the time you read this, that record may already have been broken. In just the United States, disasters in 2024 killed more than 560 people and caused nearly $183 billion in damages—more than the GDP of Arkansas.[27] Worldwide, disasters in 163 countries and territories forced people out of their homes a combined 45.8 million times, while innumerable others suffered in place.[28] No corner of the planet was safe.

The overall trend has long been clear. Each year since 2008, on average people were forced out of their homes 21.5 million times by floods, wildfires, hurricanes, and other fast-moving natural disasters, according to the UN High Commissioner for Refugees (UNHCR).[29] As in Brazil, sometimes people are uprooted, return to their home, and then must run away again just months or weeks later. Storms and floods were the cause of 97 percent of displacements globally in 2024.[30] But an absence of water can be just as dangerous as a surfeit. Just a few days after the year came to an end, wildfires ripped around the edges of Los Angeles, destroying entire neighborhoods at the center of the global entertainment industry and killing more than two dozen people. Southern California had not seen meaningful rain in months, and the amount of moisture in the soil was the lowest in a decade, inviting the fire to spread.[31] The Palisades and Eaton fires were among the most destructive in California's history and indelibly scarred the United States' second-largest city.[32]

These kinds of events are no longer extraordinary. They are now a regular part of life for large swaths of the globe. Because of human-caused climate change, the odds of disasters striking have grown higher and higher, and their devastation has grown more severe. The impacts of climate change are "widespread" and in some cases "irreversible," according to the United Nations' Intergovernmental Panel on Climate Change (IPCC), creating tremendous strain on our health, economies, infrastructure, and food and water.[33] From 2000 to 2019, climate change was responsible for global losses of nearly $2.9 trillion—about $16 million every hour.[34] By the end of this century, the planet could get on average as much as 4.4°C (8°F) warmer than the preindustrial levels of the 1800s, according to the IPCC. In even a best-case scenario, the planet will continue to heat up until the 2050s, leading to more severe weather occurring more regularly. Not every single natural disaster is the result of climate change; hurricanes, floods, fires, and droughts have been occurring for as long as this planet has been around. But the science is resoundingly clear that humankind's release of greenhouse gases has made extreme events more common, more intense, and more widespread. The consequences are disastrous for animal species all across our planet. Global warming of 2°C (3.6°F) could put as much as 18 percent of species at very high risk of extinction.[35] Mass die-offs could transform the crucial web of life that has allowed ecosystems to thrive for millennia.

For humans, the future looks hardly less severe. As many as 3.6 billion people—just under half the entire planet—live in places considered "highly vulnerable" to climate change, meaning that they are under threat from rising rivers, encroaching drought, and other impacts, according to the IPCC.[36] About a billion people live on or near the coasts, which face rising sea levels, severe storms, and floods. Warmer weather means more people will die in heat waves, while diseases such as dengue will spread to new areas, putting billions at risk of infection. Long periods of drought will create parched farmland that refuses to yield crops. In temperate climates in much of North America and Europe, roads, bridges, railways, and other infrastructure weren't built to withstand temperatures that regularly exceed 100°F. Untreated asphalt can soften, concrete can buckle, and cracks can expand to jeopardize the integrity of bolts and fixings. Air-conditioning, once a luxury limited mostly to the hottest climates, will become increasingly essential at all latitudes, straining power systems.

When these things happen, millions will leave their homes and never return. The World Bank projects that by 2050 climate change will force as many as 216 million people to move within their own countries—more than the current population of all but the world's six largest nations.[37]

This is the future. But it is also the present. Already, small island states in the Pacific are regularly inundated with water and actively planning for the likelihood that rising sea levels will wipe their homelands off the map. Croplands are going dry, and farmers are being forced to abandon their fields in search of new livelihoods somewhere else. Hurricanes routinely batter fragile rural villages, destroying homes and entire communities and scattering their residents. Worldwide, natural disasters drive people out of their homes more than twice as often as war and other conflict.[38] Even in the wealthiest country on earth, wildfires can tear through neighborhoods and floods can drown city blocks, forcing people away and preventing them from ever returning. Nearly half of Americans planning to move now say that natural disasters or extreme temperatures play some role in their decision.[39] From 2000 to 2020, more than 3.2 million people moved out of flood-risk areas of the United States.[40] And this doesn't account for the would-be migrants who were turned off places because of the climate risks. In the last few years, online real estate services such as Realtor .com and Redfin have begun to include projections about current and future climate risk in their listings; people who view these warnings on high-risk properties are more likely to turn to a lower-risk home.[41] The era of climate migration is already here. We're just finally starting to realize it.

This book is the story of this migration and displacement of people. It is happening in ways both large and small, in response to massive disasters and slow-moving environmental shifts. It affects people who live on the coasts, in dry lands, and everywhere in between. It affects rich and poor alike, although the communities most impacted tend to be farmers, herders, and others who live off the land in ways that are becoming no longer possible. By far the largest levels of climate migration are projected to be in poorer parts of the world, including sub-Saharan Africa, the Pacific, and South Asia, where many people depend on rain-fed agriculture, and which have been much less responsible for the greenhouse gases that have caused the planet to warm. Limited infrastructure in these parts of the world also makes people vulnerable to landslides and erosion, lack of access to freshwater, and other risks that high-income countries have sought to wall themselves off from.[42] The countries and companies that are most to blame for climate change, which are largely those of the Global North, have invested huge fortunes in protecting themselves from the worst impacts, at the expense of billions of others.

But climate change affects all of us. When wildfires tore through Los Angeles in early 2025, they decimated the multi-million-dollar mansions and mobile homes alike. Immediately afterward, rents in nearby neighborhoods ballooned, sometimes nearly doubling, despite a state law banning such price gouging.[43] For people who own their homes, the costs of staying in place have changed.[44] Insurance premiums on the fire-vulnerable fringes of Los Angeles and other California cities have surged dramatically, as companies come to grips with their pending financial obligations. The Los Angeles fires were likely the costliest in world history, forcing companies to pay out an estimated $40 billion.[45] In response, companies are translating abstract environmental risk into concrete dollars and cents, forcing people to move and undercutting home values. From

2021 to 2023, U.S. home insurance policy premiums increased by an average of 35 percent, with the highest increase (of 68 percent) in Florida. The cost of insurance is rising faster than inflation and growing as a share of mortgage payments, rising from around 7 percent in 2013 to more than 20 percent in 2022.[46] And that's assuming insurance is even available at all; from 2018 through 2024, insurers dropped more than 1.9 million U.S. home insurance contracts, especially in hurricane- and fire-prone areas along the coasts and in the West.[47]

Most people who leave do not go far. In low-income countries, people in the countryside often go to a nearby city; rarely do they travel internationally. Despite what certain pundits and politicians claim, crossing a border is actually very difficult. Doing so illegally is both risky and expensive. As a result, apocalyptic predictions about the number of current and future climate migrants are most likely misguided and potentially dangerous. Back in 2007, Christian Aid, a London-based development organization, predicted that as many as 1 billion people could be forced from their homes due to the impacts of climate change by 2050.[48] The Institute for Economics and Peace more recently pegged the upper estimate at 1.2 billion people.[49] Or maybe by 2100 there could be 2 billion climate migrants—about one-quarter of the current global population—another study claims.[50] Almost surely, those numbers are a gross exaggeration or rely on an extremely expansive definition of climate migrant that includes basically everyone. This book is partly designed to fight back against those narratives, which at worst feed into sensationalist fearmongering that huge numbers of immigrants from poor countries are desperately clawing at the door of the West. Even if well-intentioned, those predictions can reinforce long-running anti-immigrant tropes that foreigners are a threat to the culture, lives, and resources of the native-born. That is why this book does not use phrases like "a flood of migrants" or "a tidal wave of refugees." Floods and waves kill people, destroy buildings, and ravage communities. Human beings are not a natural disaster.

Instead, this book aims to illustrate how, as the planet reckons with an unprecedented climate shift, the way that we all live our lives is undergoing a foundational transformation. I feature here people and communities who have been impacted by climate change in multiple countries, but I only quote directly people who spoke in English; stories and words are precious things, and in cases where I relied on translation I have summarized their comments rather than inserted my own wording.

In part, theirs is an update of an old tale: Humans have always moved from one place to another in search of better opportunities and brighter futures. It is the reason I write these words from the United States rather than the homes of my ancestors in Central Europe and the Middle East. Historically, these movements have been both voluntary and forced, shaped by myriad factors including changing political winds, wars, the expansion or collapse of economic systems, and technological advances such as steamships and smartphones. Now we add to the mix climate change, a new element that is both upsetting a long-standing history of migration and compounding other drivers.

The map of human settlement is being recontoured. The lives that generations of people built over centuries will no longer be tenable. Crops will no longer grow. Homes will no longer provide safety. The ocean will not stay put. And neither will we.

1

We Have Always Been Climate Migrants

Climate migration has no single story. It doesn't even have a single definition. If you ask them, people will rarely say they are migrating solely because of climate-related factors. But dig a little deeper and there are signs that extreme and unpredictable weather are secret ingredients in a lot of the world's migration today. For instance, fewer than one in fifty migrants from West and Central Africa told researchers that they had fled their homes solely because of natural disasters or the environment, but nearly half acknowledged that environmental factors had *some* impact on their decision.[1] Often, environmental factors influence people's decision-making indirectly, such as by depressing their wages, demanding ever-more-expensive home repairs, or causing insurance premiums to rise. It is usually hard to disentangle, for instance, precisely when someone is moving because their job isn't providing enough money, because schools and other government services are underperforming, because they want to reunite with family elsewhere, or because they aren't safe—even if climate change is compounding all those issues.

Generally speaking, there are a couple of different ways to think about why, how, whether, and where people migrate. One is in terms of *push* and *pull* factors: the reasons why someone might want to leave the place they're currently living (such as a poor economy, political repression, or widespread gang violence) and the reasons why they might want to go to a new destination (perhaps a booming job market, migrant-friendly labor laws, or family connections). Climate change can affect both these sets of factors. Drought, storms, and extreme weather can make someone's current home less hospitable, while a more favorable climate or government policies that support adaptation might be a draw to go elsewhere. Another way to think about why people move is in terms of their *aspirations* and *capabilities*: what people want and how easily they can achieve it. Whether or not someone wants to move, there are an array of reasons why they might or might not be able to do so: the cost of travel, government policies that encourage or forbid migration, a social network to help them get a head start once they arrive, their ability to speak the local language, and so on. All these issues are important. Some of these factors are too big for any one individual to have control over: war, economic collapse, a flourishing job market. Some are much more personal: a friend or family member, education and skills that might allow them to thrive, or simply whether they like where they're living. Then there are a range of in-between issues: the languages people speak, or the social media platforms that show how good or bad the situation is in other places.

The researcher Robert McLeman frames the question as a sort of math problem.[2] The odds that someone migrates in response to adverse climate conditions are a function of three separate factors: their exposure to climate-related hazards (i.e., whether or not someone lives in an area that is routinely impacted by severe weather and other challenges); their sensitivity to climate issues (farmers and livestock herders, for instance, are going to be much more affected by changes in seasonal weather patterns than people who work on laptops from their couches); and their ability to adapt through methods other than migrating (such as by building dikes or rotating crops more strategically). Seen in this light, migrating is not necessarily a course of last resort, but one of several possible options. In the face of hurricanes and rising sea levels, for instance, a homeowner can board up their windows, buy expensive flood insurance, build a seawall, or decide that they'd rather live farther inland. These are all equally plausible and equally valid. If moving is cheaper and easier than the other options, so be it.

In a sense, none of this is new. Humans are an innately migratory species, and since our ancestors first stepped foot out of eastern Africa millennia ago, we have been constantly on the move. Almost always, environmental factors have played some role in the process. All major cities, for instance, sit in environmental and geographic niches that offer the basic materials for growth. Many are near rivers and other waterways to bring in goods and people, they are often adjacent to lowlands and fertile plains that at one point helped feed the growing population, and they have relatively moderate weather that allows residents to work and thrive. Even when environmental questions seem distant, they often play a hidden role in dictating where we live. My own city, Washington, DC, sits at the meeting of the Anacostia and Potomac Rivers. Its location was selected largely for political purposes, straddling the line between the early United States' politically quarrelsome North and South. Yet the emergence of distinct North and South regions, with contrasting political and economic priorities, is itself a function of the climate and environment; had its climate not been suitable for growing tobacco and other crops, the South would not have developed an economy reliant on agriculture and would not have gone to war to protect its enslavement of people who worked those fields. We live in the places we do because, years ago, someone's ancestors gathered where the conditions were suitable and then more people came to join them. The history of human geography is idiosyncratic, but it is not random. There is a reason that people do not live in Antarctica.

Environmental conditions have always changed. But historically, these changes occurred at a much slower pace (sometimes over thousands of years), only affected certain areas, and were a result of natural climatological events. In fact, scientific findings suggest that a changing climate was a crucial factor in the most critical human migration of all: the prehistoric movement out of Africa and across the planet. Our species evolved in Africa around 200,000 years ago, and we stayed there for millennia. Then, around 125,000 years ago, our ancestors may have taken their first stutter steps into Asia and Europe, after the thawing of an ancient ice age created warm, wet periods that would have spawned verdant grasses, trees, and lakes across northeastern Africa. This period of Green Sahara, as it is known, would have provided early *Homo sapiens* with plenty of food and fuel as they either trekked across the Sinai Peninsula or crossed the Red Sea into the Middle East.[3] Climate scientists' models suggest these periodic climate fluctuations recurred every 20,000 years or so, starting around 110,000 BCE, due to the earth's swaying on its axis. Each time the conditions were ripe, humans may have slowly crept out of our species' collective cradle.[4] The biggest and most important of these movements likely occurred around 50,000 to 80,000 years ago and was instrumental in populating the planet. DNA collected from people around the globe suggests that all non-Africans likely descend from people who left the continent around that time.[5]

Since that fateful prehistoric migration, climatic and environmental shifts have repeatedly pushed civilizations to expand or contract. While nothing compares to the scale and speed of climate change in our current era, the best precursor may be between about 800 CE and 1300 CE, during a period known as the Medieval Warm Period (sometimes called the Medieval Climate Anomaly). At this time, parts of the planet experienced warmer temperatures estimated to have rivaled those of the mid-twentieth century, due to a combination of increasing solar radiation (a result of changing patterns of sunspots), fewer volcanic eruptions (which spew ash, dust, and gas particles that can shade the earth from sunlight), and quirks in the earth's wobble in rotation. Like today, temperatures and conditions varied from year to year, but scientific analysis of core samples, tree rings, and other evidence shows a clear trend. For some periods and in some places—particularly in the Northern Hemisphere—during these five centuries, temperatures were as much as 1°C (1.8°F) warmer than they were in the late 1900s.[6] That may seem relatively minor, hardly enough to make you throw on a sweater much less upend a civilization. But a small temperature change can have wide-ranging repercussions. A slew of evidence suggests that around the year 1150, for instance, there were vineyards as far north as the English Midlands.[7] In the twelfth and thirteenth centuries, England's winemakers sent exports to France, causing loud complaints from continental competitors.[8]

Climate change deniers have claimed that current global

warming is simply an innocuous recurrence of this natural pattern and not a product of human-made greenhouse gas emissions. However, the Medieval Warm Period was not producing consistently warmer weather worldwide, even if its effects were widespread. Many areas such as the western North Atlantic around Baffin Bay between Canada and Greenland stayed cold, and the eastern Pacific experienced a long stretch of dry, cool conditions.[9] The phrase "Medieval Warm Period" does not mean that every place was warmer everywhere but is simply a useful shorthand to refer to a general era in which evidence suggests some areas experienced abnormally warm temperatures.[10] Also, temperatures fluctuated slowly, over the course of hundreds of years, and did not simply climb rapidly higher each and every year. This is dramatically different from today, where every single region of the globe has been affected at a never-before-seen rate.

In places where the weather was warmer, the Medieval Warm Period helped encourage migration and settlement of new areas. Ice had previously covered the sea and land in parts of Northern Europe, but the relatively mild weather of this era allowed Vikings to cross the North Atlantic and expand their colonies to the threshold of the New World.[11] To use the language of migration scholars, the warmer weather may have been a pull factor for these Vikings, promising new riches and terrain. It expanded their capabilities to travel farther by making food and fuel easier to come by. Higher temperatures and milder winters in Scandinavia may have also contributed to push factors; less frost means more robust cereal crops, which means more food, which can lead to more children, creating economic challenges if land and other resources are finite and there is not enough to go around. Norway's population increased from about 150,000 in 1000 CE to 400,000 three centuries later. A few generations of overcrowding and the prospect of hopping into a sturdy longship to settle new lands might have seemed pretty attractive.

Soon after 800 CE, Norse sailors had settled in the Orkney, Shetland, and Faroe Islands north of Scotland. A century later, they had made homes in Iceland, bringing dairy cows that, over time, would develop into a specific Icelandic breed with a relatively small body size and exceptionally colorful hide. Two generations later, Erik the Red would found the first European settlement in Greenland. His son Leif Erikson would continue west until reaching the northern tip of New-foundland, where by 1021 he or his followers founded L'Anse aux Meadows, the earliest evidence of European settlement in North America.[12] The colony is the only confirmed Norse settlement on the continent, but it was likely a base camp for further exploration. Archaeologists have uncovered mul-tiple nearby structures built out of wood and peat overlook-ing a peat bog, which are similar in style to structures built in Greenland and Iceland around the same time. Of course, slightly warmer weather was by no means the only factor en-couraging this westward expansion. But it certainly helped. It would have been harder to make the journey if the North Atlantic were entirely covered by ice or there were no grass for cattle in Iceland.

In Asia, meanwhile, evidence suggests that climate condi-tions around the same period allowed for Genghis Khan and his Mongol empire to spread across the grasslands into Korea and China, creating what was at the time the largest empire the world had ever known. The Mongols were famously horse-riding conquerors, and horses meant practically everything to them. They were transportation, weapons, wealth, and sources of food and dairy and even alcohol, by means of a fermented milk drink called *kumiss*. But horses also consume a lot of grass, and fear of overgrazing was one reason that Mongol armies on the steppes were often nomadic; staying in one place too long risked depleting that place of the fuel that allowed their armies to function.[13] It's difficult to overstate the impact that drought or abundant rains would have had on the Mongols' ability to feed and raise horses and other livestock. Overly dry conditions would have led to sparse grass and economic tensions; extra rain would have been like manna from heaven, providing an economic windfall to fund a continent-wide expansion. A study of more than a millennium of tree-ring data shows that central Mongolia underwent a dry period in parts of the 1100s, during which the society experienced political infighting and volatility, but then saw a once-in-a-millennium fifteen-year period of wetness from 1211 to 1225, just after Genghis Khan had risen to power in 1206.[14] While Genghis Khan is widely regarded as a master tactician and administrator, these conditions surely gave him a leg up in consolidating power and stretching his empire across Central Asia. While the warm, wet weather alone was not what drove his expansion across the Asian steppes, it seems to have contributed to an economic and military system that made it easier for Genghis Khan to be victorious. It was, in a sense, the perfect storm: The conditions changed to become precisely the most conducive to centralized leadership as well as to the technology and kind of warfare in which Khan's empire specialized. And he literally rode them to victory.

*　*　*

If beneficial climate conditions can be a pull factor that makes migration more attractive and allows people to expand into new areas, then adverse swings can push people away. Put another way, if a pleasant climate can encourage people to come to a new location, a harsh one can force them out. The Medieval Warm Period was followed by a five-hundred-year era known as the Little Ice Age, which, as the name suggests, meant colder weather across much of Northern Europe. The Thames River would regularly freeze during this time, and the ice around Denmark was thick enough in the 1600s to allow Danish armies to march across it into Sweden.

Returning to the Vikings, there is evidence that communities in Greenland went through severe famine in the early 1400s, around the time that the Medieval Warm Period gave way to the Little Ice Age. The cold made it harder to grow crops and raise livestock. It was also more difficult to fish and hunt for seals and walruses on the seas, which were gradually icing over.[15] The settlements didn't disappear all at once; rather, evidence suggests that smaller, rural farms were abandoned first, as people moved toward larger settlements to work for larger landowners. From there, they formed a steady migration stream back to Iceland and mainland Europe. This is a key pattern throughout history: When climate conditions change slowly and gradually turn the land barren, people don't abandon their farms all at the same time; they depart at staggered intervals, based on an array of factors such as their personal connections elsewhere, the amount of money they have saved, and whether it is safe and affordable for them to move. And they tend not to go far, most likely to the nearest town or a not-too-distant city where residents speak the same language and they might reasonably find a job plying a new trade or can push their children into a new career. Only when the entire civilization collapsed over the scale of several generations did the Norse abandon Greenland entirely.

Historically, droughts in particular have played a major role in civilizations' collapse or transformation, either by forcing large segments of the society into exodus or by rearranging the demographics and scrambling the political and economic systems that underpinned the community. This is particularly true for preindustrial, agricultural-based societies, where there is a more or less direct correlation between drought and hunger. In the American Southwest, for instance, the Medieval Warm Period was not as accommodating to expansion as it was in Europe. The period facilitated a string of droughts and temperatures that at some points were nearly as warm as those of recent decades. These conditions seem to have had a devastating impact around what is now the Four Corners area at the intersection of Arizona, Colorado, New Mexico, and Utah. Here, a society known as the Anasazi (also called the Ancient Puebloans) lived in settlements scattered across canyons and high desert. They thrived for more than one thousand years, creating a complex social and political system and building massive cities with structures as large as a modern-day city block. They may be best known today for the remarkable cliff dwellings made of stone and earth that still stand crammed into the walls of canyons and seem impossible to access without sophisticated rock-climbing equipment.

Then around the twelfth and thirteenth centuries, they suddenly disappeared. Or rather, they left. The disappearance of the Anasazi long stumped scientists, but it now seems likely the society encountered massive, decade-long droughts while its population was at its height, particularly in the late 1200s.[16]

The Anasazi were overwhelmingly reliant on growing corn; when the crops wouldn't grow in the dry weather, hunger spread, and political strife and war soon followed. There is evidence that the society went through a profound religious shift around the same time, as ceremonial structures changed from prioritizing multistory great houses to open plazas and amphitheaters. The civilization's southern fringe started to embrace the Kachina religion, elements of which continue to persist among modern Pueblo people.[17] Such a shift would seem to make sense: When food runs out, what society wouldn't start to radically question its beliefs and go searching for new gods? It is unclear if there was a specific precipitating event that occurred, but whatever the reason, large clusters of the ancient society picked up and relocated. Some went south and east to the Rio Grande area, where they could depend on the river's cool water to irrigate their crops. As they traveled, the Anasazi culture died off or melded with those of other communities along the way.

Meanwhile, a bit farther south, in the Yucatán Peninsula, the Mayan civilization underwent a severe political crisis around 900, resulting in the depopulation of many major cities. Scholars still don't understand precisely what happened, but analysis of climatological evidence such as layers of lake bed sediment suggests that one factor may have been intense droughts in the ninth and early tenth centuries.[18] The civilization had previously expanded in wet, lush periods, but it strained during the drought, which seems to have deeply affected its political and economic structure. The Maya had a complex system of reservoirs for water management, which became useless when the rains dried up. To be clear, the society did not simply die off from thirst. But it seems possible that drought led to hunger and disease, and just as importantly undermined public faith in rulers' ability to protect and care for them. This cascade of factors undermined support for the central political leadership, spurring political conflict and unrest and ultimately leading people to flee major cities.[19]

In other eras, scholars have linked drought to the decline of Mesopotamia's ancient Akkadian Empire, the Bronze Age Harappan civilization in the Indus Valley, and Egypt's Old Kingdom, among others. Intense monsoons following a prolonged drought in the city of Angkor, the capital of Southeast Asia's Khmer empire, led to its abandonment in the early fifteenth century. Drought is special in this way. It is a classic example of a slow-onset type of climate impact that might not necessarily alter civilizations in a year or two, but when it stretches for multiple years—or multiple decades—can be devastating. Even today, in our era of indoor plumbing and modern hydrology, drought continues to be serious. Overall, lack of water results in five times as much migration as an excess of water, according to a team of World Bank researchers.[20]

The history of climate and migration is also one of disease. A warmer, wetter climate allows rodents, mites, and other creatures carrying bacteria and viruses to spread alongside—and on the backs of—humans. As these creatures intermingle, germs and viruses inevitably jump from animal to human, potentially causing massive depopulation as people die, flee, and stay away from infected areas.

A prime example is the Black Death in the Middle Ages, a bubonic plague that swept from Asia across Europe in the mid-fourteenth century and killed more than 25 million people—about one-third of Europe. The Black Death was most likely a by-product of climate-driven migration and, in its own way, was also a driver of additional migration. Probably, the fleas carrying the plague bacterium (*Yersinia pestis*), which famously spread on the backs of rats, were ushered into Europe because of climate fluctuations in the northern Asian steppes. A few years of plentiful rain led to a large population of plague-carrying rats, which then began to starve as the seasons changed and the land became less habitable. Desperate for food, the rats extended their range south, making contact with other rodents and the humans who lived nearby, and forcing fleas onto new hosts.[21] Some research suggests that fluctuating climate conditions in Asia may have led to a steady, pulsing influx of new bacterium strains from Asia, as fleas carrying *Yersinia pestis* were continually reintroduced.[22] After its initial introduction in Europe in 1347, the plague returned again and again over the course of three hundred years; every generation was hit with a new outbreak.

The plague also had profound long-term consequences for the structure of societies, irrevocably altering Western notions of work, culture, and livelihoods and causing high levels of migration as people sought out new work and moved to towns.[23] Depopulation due to the large-scale death dramatically raised peasants' bargaining power, allowing them to demand more money and significantly raising their standard of living. For maybe the first time, peasants in places such as the United Kingdom who were getting squeezed had a choice: If conditions were pitiful, they could simply go somewhere else offering better pay. Fewer people meant there was comparatively more farmland to go around, and gradually the system pivoted from serfdom to something like wage labor. "Such a shortage of workers ensued that the humble turned up their noses at employment, and could scarcely be persuaded to serve the eminent unless for triple wages," a cathedral chronicle in the English town of Rochester claimed in the mid-1300s. "As a result, churchmen, knights, and other worthies have been forced to thresh their corn, plough the land and perform every other unskilled task if they are to make their own bread."[24] As peasants began wielding their newfound bargaining power, they flocked to places where they could secure a more comfortable remuneration for their labor. Eventually, serfdom and the feudal system fell apart.

The demand for labor also impacted the Vikings, who at the time were at the tail end of their forays into Greenland and the New World. With pressing economic needs back on the continent, many Vikings had new incentives to return to the mainland, potentially helping to stymie their further expansion.

There is a danger to attributing too much to climate change, and I want to avoid the suggestion that a little rain or a dry spell was the sole or even primary reason why civilizations rose and fell. Pleasant weather doesn't guarantee that people will move in, just as hazardous conditions don't guarantee that people will flee. In each of these historical cases, changes in climate and weather were only part of the story. There were also wars and political contests and economic booms and busts and technological advancements that impacted the likelihood that a city and civilization would rise and fall. In Greenland, for instance, Norse communities reliant on trading walrus ivory also contended with competitors from Russia and Africa, devaluing their product and creating economic headwinds.[25] If any of a range of factors had been radically different, the outcome might have changed. My point here is only to show that the climate and environment have played significant roles in human migration over millennia. Humans are complex beings with a range of needs, and we move or don't for a million reasons. It can be dangerous to suggest that any single issue is what's driving us, or that we respond to changes in the seasons like puppets on strings.

Moreover, migration is just one of many possible responses to drought, disease, and other difficult conditions. Historically, some people also starved, while others adjusted their lives and invented new ways of muddling along. Moving is rarely the first step. And when people do migrate, even on the scale of large societies, rarely does it start out as permanent. Various segments of Anasazi society, for instance, constantly moved around during multiple periods of drought. In the mid-1100s, well before the entire society collapsed, a fifty-year drought in what is now northwest New Mexico prompted many people to flee north to the Mesa Verde area of southwestern Colorado, where they helped form larger and more complex cities.[26] In doing so, they put further strain on the environment, cutting down more trees for buildings and fuel and diminishing the population of wild game, which would make it harder for the society to withstand another drought over the next 150 years. Moving to nearby cities is a classic strategy for people encountering adverse climate conditions. Especially in modern times, cities offer jobs, stability, and support networks that rural areas might not. But as the Mesa Verde case shows, in the wrong conditions this influx may simply aggravate the situation; more people means more strain on resources, which might only make matters worse as conditions continue to deteriorate.

Finally, many of these events occurred so long ago that it is nearly impossible to reconstruct exactly what happened. Historians have a hard enough time analyzing events that were meticulously recorded; trying to tell a societal narrative by relying on ice cores, sediment samples, tree rings, ruins of ancient buildings, and other historical detritus from a thousand years ago is like trying to put together a jigsaw puzzle with most of the pieces twisted and all of the image rubbed off. No one alive today was there to experience how or why these civilizations went through dramatic changes, and the past is too far gone to know for sure. Among scholars, these narratives are still topics of constant debate, and almost surely some expert will take issue with the stories as I have presented them here. Moreover, the scientific evidence provides only a thirty-thousand-foot look at the kinds of opportunities and stresses that societies face when the climate changes. Individuals act in discrete and often confusing ways, motivated by their love for their families, their desire to get a good job, their belief in their government and their god, and a million other influences.

For a clearer understanding of how this plays out, we can turn to more a recent example on which there is contemporaneous reporting and documentation that offer a wealth of insight into precisely how changing environmental conditions caused people to migrate. One that might be best known to Americans is the Dust Bowl, the period of U.S. and Canadian history from around 1930 to 1940, when a series of severe multiyear droughts hit the Southern Great Plains states and devastated millions of acres of farmland. As perhaps best captured in John Steinbeck's classic novel *The Grapes of Wrath* or the evocative and poignant photography of Dorothea Lange, about 2.5 million people abandoned their homes in Arkansas, Oklahoma, Texas, and other states, many heading to California, where the weather was fairer and jobs were believed to be readily available (although as Steinbeck illustrates, migrants often found themselves in just as precarious a situation as the one they had left in the Great Plains). By 1950, former residents of Oklahoma, Texas, Arkansas, and Missouri accounted for about three out of every twenty-five people in California.[27] In short: A drought and dust storms hit in several waves in the midst of the Great Depression, compounding economic challenges for ranchers and farmers who were already eking out a living and struggling to keep up with the country's rapid lurch into twentieth-century industrialization. We don't have to make inferences from core samples to understand what happened in the Dust Bowl; we can get the story straight from the migrants themselves, who were known as Okies and demonstrated firsthand what environment-driven migration looked like even in the pre–climate change era.

Migration to California had been going on for years before the Dust Bowl. Drought, too, was nothing new, and in fact the drought of the 1930s was probably shorter and less severe than droughts of previous centuries or droughts that have occurred since.[28] The region suffered an equally severe drought in 2011 and 2012, which devastated crops and pastures across the region. But the Dust Bowl–era drought was particularly expansive, stretching from roughly Texas's border with Mexico into the Canadian provinces of Alberta and Saskatchewan.[29] Technically, the dust storms and wind erosion that most vividly captured the public imagination were responsible for only a tiny share of the Dust Bowl–era migration. Fewer than 16,000 people from the teardrop-shaped region that experienced the dust storms—up from the panhandle of Texas into New Mexico, Colorado, and Kansas—ended up in California, out of the nearly 1.4 million natives of the western South to be in the state as of 1950.[30] But for those from this area, the drought's impact was clear: nearly 10 percent of residents of counties suffering from high levels of erosion migrated more than two hundred miles between 1935 and 1940, a rate almost 3 percent higher than in other counties.[31]

What made this drought so impactful is that it occurred at a pivotal time, right in the depths of the Great Depression and amid a landmark shift in agricultural production methods. Weather conditions aside, the 1930s were a hard time to be a farmer. As the economy plummeted amid the Depression, markets for agricultural products dried up. A pound of cotton that would fetch 16¢ in 1929 was worth less than 7¢ in 1932.[32] The 1933 Agricultural Adjustment Act offered cash to farmers to take land out of production, benefiting farmers who owned their own land but not the tenant farmers who worked more than half of farms in Oklahoma, Arkansas, and Texas, and who were evicted when the owner of their land took a payout and sold it out from under them. Sometimes, landlord farmers used the AAA money to buy tractors and other farming equipment that made tenant farmers redundant, effectively replacing them with machines and leaving them unemployed.

This crash was particularly dire because of its contrast with the promises of just a few years earlier. World War I had triggered high commodity prices and, after the war ended, farming seemed like a can't-lose proposition. In the 1920s, people raced to get into agriculture, sure that they were about to hit the jackpot. This "great plow-up" created a new generation of farmers with little experience in the industry and little understanding of how to appropriately manage the land.[33] Some were so-called suitcase farmers, who lived far from their fields and popped in only to plant and harvest their crops. It was a risky bet they were taking—that farms would be productive even with little day-to-day care—which worked out well when the weather was good but failed spectacularly when the rains stopped.[34]

Around this time, too, motorized tractors began to replace farm animals for pulling plows, and systems of finance and credit grew more complex. That gave many farmers the ability to finance extra power for their fields but also plunged many into debt to pay for technologies they could not afford.[35] Farming practices of the day also likely contributed to soil erosion, which led to dust storms, meaning that agricultural practices had negative economic and environmental effects of their own. Many farmers resisted irrigation, believing that the otherwise typically wet environment would keep their crops well-fed. Their plows also tilled up a surface layer of soil that became loose and spread easily in the wind.[36]

And then there was the boll weevil, a beetle just a quarter inch long that feeds on cotton plants and spread throughout cotton-producing regions of the United States in the 1920s. Infestations were a "wave of evil," U.S. Department of Agriculture official B.T. Galloway testified to Congress in 1903.[37] In other parts of the country, boll weevil infestations would be a major impetus for the Great Migration of Black Americans from the South to the North, Midwest, and West. (Ninety-five percent of Dust Bowl–era migrants were white; whereas Black residents of these states had been leaving in sizable numbers throughout the early twentieth century, they mostly headed north and east at this time.) Accounting for inflation, the little beetle was responsible for $100 billion in losses in the United States alone during the twentieth century, according to the USDA.[38] Not until the 1970s, when it was the target of one-third of all pesticides applied in the United States, would the government develop a system to get rid of the pest.

Was there a final push that convinced people to head to California? Often, people found themselves in the midst of a broader wave of migration and got caught up in the momentum. It's one thing to be the first one to leave your hometown and strike out for something new. But it's something else entirely to follow suit after everyone else in town has already left, leaving behind shuttered shops, decaying houses, and even fewer jobs than before. At least half of migrants traveling in the 1930s were following relatives who had already settled in California, and in some places as many as two-thirds of migrants were following earlier generations.[39] These connections might tell them where to go, help them get a new job or find a place to stay, and ease their settling in.

Not everyone wanted to go or was able to. The Great Plains were home, after all, and in general people like their home and don't want to leave. In better times, down-on-their-luck farmers might simply have gone into town to find a job at a factory or rail yard, but in the depths of the Depression unemployment escalated everywhere, and newcomers from the sticks were hard-pressed to fill a job that local city slickers couldn't fill first. Still, many tried all the same, going to nearby towns and cities to ply a new trade as they could. Only afterward, when things didn't work out, did they head west.

Indeed, many migrants who made the long trek to California did so only after exhausting other opportunities closer to home. Given the itinerant nature of farmwork at the time, many agricultural workers would have moved several times, including into and out of cities. According to one survey, 36 percent of Oklahoma farmers in 1937 had previously spent some amount of time working outside of agriculture.[40] In that light, moving westward was not a radical break from an otherwise sedentary lifestyle, but just one especially long relocation in a life that had been checkered by them.

Dust Bowl migrants shared other qualities common with migrants of all eras, namely, that they were people who could afford to travel and understood the opportunities for doing so. Their aspirations, in other words, were met by their capabilities. Most were young, typically in their twenties and thirties. They tended to be slightly better educated than those who stayed behind. A slight majority (53 percent) were men, although many went as families.[41]

They also benefited from tremendous technological advances and infrastructural investment—reliable automobiles and a well-traveled road—that quite literally paved their way. Just a generation earlier, a journey of more than 1,500 miles would have been unthinkable, but cheaper cars and roads such as the iconic Route 66 made the drive possible in just three days. And if things didn't work out, they could come back— so long as they had enough money for a few tanks of gas. In fact, some migrants went back and forth regularly, in a pattern commonly referred to as circular migration. Also important was the presence of cheap printing and photography, which at first made it easy for California landowners to publish lavish advertising pamphlets promising would-be migrants how great their new homes would be.

The Dust Bowl migration occurred solely within the United

States, so migrants did not need to worry about things like visas and documentation in order to live in California legally. At least, not officially. In reality, migrants often faced pressure from authorities and locals who saw newcomers as competition driving down wages. Often, migrants were shunned. While they tended to have more money than the people they left behind, they usually spent a lot to make the move (typically believing they would cash in upon arrival) and were thereby often poorer than locals. They were also by nature itinerant, which often caused distrust. "The comfortable people in tight houses felt pity at first, and then distaste, and finally hatred for the migrant people," writes Steinbeck. California authorities also vocally tried to discourage migration, but the message stood little chance in competition with the idea of the state as a land of plenty.

After arrival, Okies who had dreamed of living off the fat of the California land mostly came up skinny. Wages were low, newcomers were looked down upon, and the prospects dimmed for each new wave of arrivals. Moreover, the unsteady pace of seasonal agricultural meant that workers experienced a few short weeks or months of flush times but then often were confronted with several months of unemployment. By 1940, only 2 percent of recent migrants to the San Joaquin Valley owned their own farm or were managers or tenants, and those who arrived later tended to have worse outcomes. Some cut their losses and headed back east, where at least they knew other people and had a bit of a safety net to rely on.

Often, new arrivals faced a scarcity of available housing (a problem that continues to plague California today). And so, drawing on the industriousness that had brought them to the Golden State in the first place, many built their own accommodation. Typically makeshift, houses in subdivisions known as "Little Oklahomas" began to spring up on the outskirts of cities across the agricultural region, with sizable neighborhoods outside of Bakersfield, Modesto, Salinas, Fresno, Stockton, and Sacramento. Land was relatively cheap, but the lack of regulations and building codes meant areas often lacked adequate sanitation. Eventually, neighborhoods like these would give birth to a new generation of rural Californians. Okies who went to cities faced similar challenges but were eventually integrated into the larger urban culture. By 1940, most newly settled people in Los Angeles and other similar urban centers had found a job and had secured a standard of living greater than what they had left behind.[42]

Government programs helped. Dozens of labor camps were operated by the Farm Security Administration throughout California, which provided shelter to tens of thousands of people. The camps typically offered running water, bathrooms, and other facilities that were welcome relief. But FSA rules limited residence to one year, in an effort to ensure the camps stayed as temporary emergency shelters and did not become permanent residences. The government also invested in efforts to reduce erosion, manage the soil, and otherwise adapt to the environmental changes in the Plains states, in part by offering farmers subsidies and buying up land itself. Federal agencies also began investigating long-range weather forecasting, offered farmers loans and subsidized animal feed, and even shipped food to some of the hardest-hit areas. This assistance was never comprehensively tracked but has been estimated to have reached as high as $1 billion in 1930 dollars (the equivalent of nearly $19 billion today).[43]

*   *   *

The Dust Bowl did not spell the end of the Great Plains. Eventually, communities and the country recovered from the drought and the Depression and built new industries. Some migrants returned, some stayed connected to their old lands from afar, and some cut ties entirely. Their stories are not those of the Vikings traveling from the Faroe Islands to Newfoundland, or of the Anasazi across whose ancestral home many Okies drove on their journey to California. Their societies grew and evolved, as all societies do. All of us are formed by migrations of previous eras. Whether driven by penalty of death, promise of opportunity, or simple curiosity, migration is a natural human instinct, and it is one that is inextricably linked with the story of the natural world. Yet as history shows, many societies are not always so lucky as California. The legacy of the Anasazi is most visible in the abandoned ruins left behind. The Norse colony of L'Anse aux Meadows was deserted not long after it was created, and now it is an out-of-the-way tourist stop at the far tip of North America, a gift shop selling T-shirts and postcards.

Most of these historical examples were driven by drought, a killer for agricultural societies. While still a significant driver of migration, especially from rural areas, drought in the post-industrial age is no longer the existential threat that it once was. Famines are, thankfully, exceedingly rare, and now mostly a by-product of political failures or war crimes. In most of the world, societies are not erased by fields that go barren. Instead, the problem is precisely the opposite: Some of us are drowning.

2

Preparing for a New Atlantis

The cerulean expanse of the Pacific Ocean occupies nearly one-third of our planet. It is dotted intermittently with low-lying islands that look idyllic from the white-sand shore. Palm trees lean idly into warm turquoise waters in which tropical fish dart through coral and mangroves. Many small islands are in strings of atolls: the flat, ring-shaped tips of underwater volcanoes on which mounds of coral have built up just enough to peek over the waves, creating strips of low land circling a calm lagoon. The dry land usually rests in broken segments barely a football field wide. From above, they appear to form a series of dotted lines in the water, colored alternatingly by pale earth and green treetops. The Carteret Islands. Fiji. Kiribati. The Maldives. The Marshall Islands. Tuvalu. They pepper the vast seascape that most maps consider to be empty, the place cartographers cut in half to serve as both the planet's beginning and end—an empty nothingness to bookend the rest of the world.

Many of these islands are sinking. Or more accurately, the sea is rising to swallow them. By the end of this century, the United Nations' Intergovernmental Panel on Climate Change (IPCC) warns that sea levels may be several feet higher than today, rendering some of these nations entirely uninhabitable.[1] Millions of people live here, at least for now. But the ocean is submerging the land. Homes, cultures, and histories may be lost forever. One-third of the fifteen most climate-vulnerable countries in the world are located in the Pacific.[2] If the seas continue to rise, the only options are to leave or drown. The problem is both undeniable and immediate. In response, communities and governments are making difficult calculations about how, where, and when to move.

The engulfment has already begun. From 1947 to 2014, five of the more than one thousand low-lying land masses that comprise the Melanesian country of the Solomon Islands, east of Papua New Guinea, simply vanished beneath the waves.[3] "What we are seeing there will become the norm," Simon Albert, a researcher who studied the disappearing islands, told CNN.[4] In 2018, the remote East Island in northwestern Hawaii, the far-north Perlamutrovy Island off the Arctic coast of Russia, and a tiny rock of an island off the coast of Japan all disappeared, wiped off the map by the ocean's slow but steady rise.[5] These already-gone islands were uninhabited, barely perceptible outcrops of stone and earth in the middle of the water. They were mere rounding errors in the earth's geological history. But in the dramatization of our climatic journey, they are the harbingers of a future beneath the waves.

Take, for instance, Kiribati, a country of over 100,000 people scattered across dozens of islands halfway between South America and Australia. Kiribati (pronounced *keer-uh-bas*) is the only country in the world to sit in all four hemispheres (Eastern, Western, Northern, and Southern) and is sometimes remembered by outsiders for its role in World War II, when U.S. forces invaded the Japanese-held islands. These days, it sits firmly on the front lines of the climate threat. In most parts of the country, you will be no more than six or seven feet above sea level. More than four-fifths of locals say they have been directly affected by sea-level rise, and it is hard to find a spot on the islands where the sea is not either already encroaching onto land or threatening to do so.[6] Beaches have disappeared, homes have been abandoned, and during high tides waves wash over the capital's main causeway.[7] In 1999, the islets of Tebua Tarawa and Abanuea (a name that ironically translates to "the beach that is long-lasting") disappeared underwater.[8] At the end of this century, the nation's entire landmass is likely to be submerged, at which point it will become the first country ended not by war or revolution but by climate change. Homes and towns that stood for generations will simply cease to be.

As in the Dust Bowl, movement out of Kiribati is a long, slow process. Populations do not disappear in a rush but rather make a series of small separate decisions that add up to emigration. The process is well underway. Already, many i-Kiribati people (as natives are known) have gone to the capital atoll, Tarawa, a collection of thin islets falling in a sideways V-shape, resembling the mathematical symbol for greater than (>). Although Kiribati's population is not large, it is remarkably dense. About half of the total country lives in Tarawa, where some areas have twice the population density of Tokyo: equivalent to about 38,000 inhabitants per square mile.[9] The beach is crowded with open shacks just a few feet wide, many belonging to newcomers from outlying islands.

There are complicated logistical and legal questions that will arise when and if countries like Kiribati disappear for good. If a country is wiped off the map, does its government similarly cease to exist? Do its thousands of residents become stateless? Does a sunken country lose its seat at the United Nations? Possibly, but not necessarily, legal scholars suggest, although it's never really happened—at least not in this way—so no one really knows.[10] Tuvalu, which sits about eight hundred miles south of Kiribati, has already begun to plan for this eventuality. Since 2023, the country's constitution has been amended to ensure that "the State of Tuvalu within its historical, cultural and legal framework shall remain in perpetuity in the future, notwithstanding the impacts of climate change or other causes resulting in loss to the physical territory of Tuvalu."[11] In other words, even if climate change wins the battle over Tuvalu's soil, the country itself will persist.

Entire countries don't have to disappear to threaten people's lives. An estimated 90 percent of Pacific Islanders live within a few thousand feet of the coast, and most have more than half their infrastructure within just a couple of city blocks of the shoreline.[12] The problems in the Pacific may be more clear-cut than elsewhere, but the rising oceans are a problem for all of us. Worldwide, almost 1 billion people—one-eighth of the global population—live within about six miles of a coastline, and about 2.75 billion live within sixty miles of the coast; because of the planet's rapid urbanization, numbers are likely to rise significantly by the end of the century.[13]

Our planet is mostly water, and throughout human history we have relied on the ocean for food, trade, and transportation. Now the sea is turning on us. Precisely how much and how quickly is a matter of debate and depends in large part on how and whether the planet is able to stop heating up. The IPCC says the seas could climb by more than three feet by the end of the century, but some scientists consider that to be a conservative estimate. One more extreme prediction, which relies on the exponential melting of the Antarctic ice sheet, predicts that the planet could see as much as nine feet of sea-level rise by 2100.[14] The Arctic and Antarctic are particularly mercurial, in that they create something of a feedback loop: White ice reflects as much as 90 percent of the sun's light and heat back into space, but when that cold white ice melts and gives way to inky black ocean, the reflectivity (or albedo) changes, and ten or fifteen times as much heat energy is absorbed. Melting ice begets melting ice, creating a feedback loop. Over the last twenty years, the Arctic has warmed more than twice as fast as the rest of the world.[15] All that melting ice goes right into the seas, where it gradually pushes up the level of our globe's big bathtub, splashing water out the sides and into our homes.

In the history of our planet, seas have risen and fallen regularly, although the time scales are huge. About 500 million years ago, global sea levels were much higher than they are now, and they have fluctuated repeatedly in the eons since. The earth's wobble, the tilt of its axis, and the not-quite-precisely-circular shape of its orbit around the sun mean that every 100,000 years or so the global temperature drops just enough to cause an ice age. The amount of water on the planet is finite, but global temperatures determine how much of it appears in its liquid form versus as ice. During the last ice age, about 21,000 years ago, there was roughly two and a half times as much ice on land as there is today, and consequently sea levels were much lower.[16] This was the era of the land bridge across the Bering Strait, when humans could walk from what is now Russia to Alaska and begin to populate the Americas. Before that, about 125,000 years ago, was the Eemian period, when the world was a little bit warmer than it is now, and the seas were about 26 feet higher. If you melted all the world's glaciers and ice caps, turning every bit of ice into liquid water, sea levels would be about 212 feet higher than they are currently.[17] Based on historical patterns and planetary logic, the globe ought to be passing through its warm era and heading into a new cooling phase right about now, but human-caused climate change has switched us back into warming mode. Sea levels rose about eight inches over the last century, but the rate is accelerating. Warmer water also has a slightly larger volume than colder water, so as the oceans heat up they also expand and push the shoreline ever upward. This process is responsible for about half the sea-level rise observed over the last fifty years.[18]

Some of these changes are already locked in. Even if we got our carbon emissions under control tomorrow, there is a lag between warmer oceans and melting ice, meaning the seas will continue to rise for at least the next few decades no matter what mitigation measures we take. It is already too late.

The seas are rising particularly fast in many of the places where humans live. Despite what the notion of being "at sea level" may suggest, the seas are not actually entirely level. Ocean currents linked to warm and cold fronts that move around the planet create imbalances in the sea, moving water from one coast to another. As the Gulf Stream—which controls temperatures in parts of the North Atlantic by bringing cold water down to the equator and warm water back up—slows, it creates hills of water toward the coasts, bringing higher tides. Differences in air temperature create wind patterns that do the same thing, pushing around the tops of waves similar to the creation of sand dunes in the Sahara. Continents also rise and sink slightly—usually just fractions of an inch per year—based in part on how heavy they are. When glaciers melt, they reduce the weight on the underlying landmasses, causing them to rise slightly in the water. Also, the planet's gravity is slightly uneven; glaciers (like all objects) create gravity, which draws tiny swells of water nearer to them. As glaciers lose mass, they create less gravity, meaning that they attract a fractionally smaller amount of nearby water. That water has to go somewhere, so oceans on the other side of the planet get a tiny bit higher in response.

As a result, the seas are actually rising quicker in places far away from the melting glaciers, which is precisely where most of us live. For instance, sea-level rise linked to the melting of the Antarctic ice is 52 percent faster in California and Florida than the global average, even though Antarctica is eight thousand miles away.[19] The seas are also rising particularly fast in the Pacific—two or three times faster than the global average.[20]

For people, the risks are multiple. Rising sea levels mean a greater risk of flooding, as tides reach higher land and magnify tropical storm surge. As the waves lap at the sand and then retreat, they often drag the coast with them, causing erosion that eats the ground out from underneath buildings and towns. Even inland communities may be at risk; rising oceans can seep into groundwater, polluting freshwater with salt and making it impossible to grow crops or pull drinking water from the earth.[21]

When the sea comes knocking, one option is to adapt, such as by putting houses on stilts or building seawalls and dikes to keep nature at bay. Historically, those measures have been tremendously important for communities around the world. Much of the Netherlands sits below sea level, for instance, but it has relied on systems of barriers and pumps to keep the

North Sea at arm's reach. Similarly, groundwater stays close to the surface in lake-surrounded New Orleans, so the dead tend to be buried above ground. There are a variety of ways to manage the encroaching waters, and we as a species have gotten pretty good at them.

Communities in the Pacific are trying to follow suit. In late 2023, Tuvalu's capital atoll, Funafuti, completed a dredging project to raise the island above the seas, or at least part of it. The island is only a few feet above sea level, but a nearly $40 million effort reclaimed land to act as a buffer between the land and the ocean and should help protect the island at least until the end of the century.[22]

But defenses don't work everywhere. Many tropical islands rest on top of the remnants of ancient coral reefs that have collapsed and turned to limestone, a porous rock that allows water to seep through. Even if communities were to build seawalls in these areas, the water could simply come up through the ground as if through a sieve. Seawalls might also backfire, if they are built in such a way that simply traps storm surge close to land rather than keeping it farther out to sea. And they can make erosion worse, by preventing efforts to replace the sand dragged out by the tide from the bottom of the wall.

Moreover, these kinds of defensive measures are extremely expensive, take a very long time to build, and tend to be ugly. The $40 million spent on the project in Tuvalu amounts to about two-thirds of the country's entire annual economic output and will protect only part of one of its nine islands. A $6 billion proposal to build a twenty-foot-tall sea wall in Miami, the U.S. city that is simultaneously most threatened by sea-level rise and arguably the most image-conscious, met opposition from residents who worried about a concrete eyesore in Biscayne Bay, adjacent to the gleaming financial district. Venice, Italy, spent only slightly less money and multiple decades to build an intricate system of barriers that rise and lower as needed to protect the Venetian Lagoon from surges during exceptional storms. MOSE, an acronym for Modulo Sperimentale Elettromeccanico (Experimental Electromechanical Model) that evokes the biblical Moses, went into operation in 2020 and has become a crucial crutch for the iconic city in the few years since. New York City broke ground in 2022 on a multiphase, $2.7 billion project to protect lower Manhattan with a system known as the Big U. The project was launched after 2012, when Hurricane Sandy flooded nearly one-fifth of the city and wreaked $19 billion of damage, and seeks to encase the southern tip of the island with raised parkland, flood walls, flip-up barriers, and other resilience measures.[23] By the end of the century, governments collectively may spend nearly $100 billion on protections for the world's coastlines.[24] And costs are only going up, since the longer we put off building protections, the more expensive they become.

Not all places are Venice or New York, however. Political leaders make choices about which places get multibillion-dollar sea defenses and which places are left exposed. How much is worth saving? And who gets to decide? Let's not be coy. Wealthy former colonial powers in Europe and North America are going to be the major beneficiaries of expensive and time-consuming infrastructure projects. Within those countries, financial districts, tourist destinations, and other high-profile sites are going to be at the top of the list. Lower-income, predominantly non-white communities are going to be left behind. It's a matter of financial costs and benefits, and the people and governments with money—the ones most responsible for climate change—decided long ago to protect themselves first.

The Marshall Islands, thin dots of land on a vast stretch of sea one thousand miles northwest of Tuvalu, has estimated a need of about $35 billion to raise some land by more than twelve feet and dredge the ocean floor to build new reclaimed land.[25] That adds up to more than a hundred times the country's annual economic output. If the money does not come through, then the country is preparing for the harsh future in which it no longer exists. "If by 2100, no decision can be made to protect areas of atolls to the [six-foot] sea level rise level, or if there is no funding for it," the country says in its climate adaptation plan, "then the decision must be to help all population to migrate away from [the Republic of the Marshall Islands]."[26]

Without adequate defenses from the rising sea, a contingency plan like the one proposed by the Marshall Islands is really the only choice available: Move to higher ground. The number of people who will move in the next couple of decades is unknowable. It could be millions, or tens of millions, or hundreds of millions by the end of the century, depending on how much the seas rise, what investments are made to keep which people in place, and how government policies encourage or block movement (such as by offering home buyouts in flood-threatened zones or making it legal to move internationally).

For low-lying Pacific Island communities and nations, the decision is nothing less than existential. Without protection, people will leave. They cannot go to higher ground, because many of these islands simply do not have any. "In Tuvalu and Kiribati, it's very difficult to move communities to safer, more inland places, because basically the very small islands are so narrow that there is no inland," explained Carol Farbotko, an Australian researcher who focuses on Pacific Island communities. In Tuvalu, for instance, "the first choice of the government and much of the population is to stay in Tuvalu as long as possible. So any form of wholesale population relocation is considered very much a last resort. But that last resort option is . . . definitely being taken seriously," she told me.[27] Some countries have islands that are slightly larger, higher, and better defended against the sea, which might offer promise. Or else people will look to go abroad. Either choice has its pros and cons, its costs and benefits. There are no good answers.

Leaders have begun weighing these options and planning for the day when that move is inevitable. Their plans involve not just constitutional amendments to ensure the survival of Tuvalu's sovereignty even if its land disappears, but also nitty-gritty details to help individuals get from Point A to Point B, to build new homes, find new jobs, and create new schools in new countries. In 2014, Kiribati paid nearly $9 million to buy eight square miles of land in Fiji, in case i-Kiribati people need to build a new home. "We would hope not to put everyone on [this] one piece of land, but if it became absolutely necessary, yes, we could do it," President Anote Tong said at the time.[28] The Maldives has used tourist revenue to establish a sovereign wealth fund that one day could be used to buy land for people in Australia, India, or Sri Lanka.[29]

Researchers and policymakers call these approaches a form of managed retreat, or planned relocation. At their heart is the notion of purposefully picking up an entire community and moving it someplace else. On a much smaller scale, planned relocation has occurred hundreds of times in all corners of the globe for more than a century. Many relocations may seem minor—worldwide, most entail moving only a few hundred households less than a mile and a half from their initial location.[30] The U.S. government has done it repeatedly, offering buyouts and other incentives to encourage people to leave flood-prone areas; more than half stay within ten miles of their original site.[31]

One of the earliest examples was Niobrara, Nebraska, a small village of a few hundred people at the confluence of the Missouri and Niobrara Rivers, just over the border from South Dakota and an hour and a half west of Sioux City, Iowa. These days, Niobrara is essentially a couple of streets off the intersection of Highway 12 and Highway 14. There's a gas station or two, a diner, a bar with an old-timey saloon facade, and not much else. The town was founded in 1857 with hopes of assisting steamboats moving down the Missouri and quickly filled up with a handful of general stores, a blacksmith, and a three-story hotel that was at the time the largest in the state.[32] But a few decades later, in the so-called Hard Winter of 1880–81, a block of ice formed a natural dam downstream on the Missouri, backing up the river and causing floods for one hundred miles upstream.[33] Within an hour, Niobrara was inundated in up to six feet of water, which did not recede for a week. Photos from the era show frigid water up to the eaves of houses and barns, spilling into windows. Chunks of ice settled in the streets. Buildings that did not collapse were soaked through.[34] In response, the town decided to pick up and move. Not far— just a mile and a half or so to the southwest, slightly up a hill where the rivers seemed less likely to invade. Only a couple of weeks after the flood, entire houses had been uprooted, placed on trailers, and hauled uphill by teams of oxen, mules, and horses.[35] Most of the entire town was relocated by the time 1882 rolled around.

Niobrara's experience is notable not just for this one move more than 150 years ago but because it moved again, earning it the nickname "the town too tough to stay put." This time the culprit was the construction of the Gavins Point Dam about thirty-five miles downstream, which backed up water to create Lewis and Clark Lake (the Lewis and Clark Expedition had camped nearby while it traced the Missouri for this part of its early nineteenth-century expedition). That blocked sediment coming into the Missouri from the Niobrara River, settled tons of sand and mud on the lake bed, and gradually raised the water level upstream. The river began creeping into the cellars and ditches of the village of Niobrara, transforming it into a mosquito- and cattail-ridden swamp.[36] So in 1971, ninety years after the initial move, the residents of Niobrara voted overwhelmingly to move again, following an Army Corps of Engineers plan to relocate uphill once more. This move was a bit more protracted than the first. It took nearly five years, and many residents either chose not to move or went elsewhere.[37] Locals received cash buyouts based on "fair market value" of their property, which they said were inadequate.[38] Still, it was incredibly expensive: Niobrara's initial move cost $40,000 (about $1.3 million in today's money); the sequel in the 1970s cost about $14.5 million (about $115 million today).[39]

Over the last century, small moves like this have become increasingly common, though they are not always billed as acts of migration. In 1937, the catastrophic flooding of the

Ohio River prompted Leavenworth, Indiana, and Shawnee-
town, Illinois, to be rebuilt off the riverbanks and up on higher
ground. A generation later, half the town of Soldiers Grove,
Wisconsin, was relocated off the Kickapoo River floodplain.
And after the so-called Great Flood of the Mississippi and
Missouri Rivers in 1993, the towns of Valmeyer, Illinois, and
Rhineland and Pattonsburg, Missouri, relocated.[40] In 2016,
the Obama administration provided more than $48 million
to relocate the community of Isle de Jean Charles, Louisiana,
a barely-there mound of green in the liminal space where the
land disappears into the Gulf of Mexico. The island had once
stretched for more than 22,000 acres, but more than 98 percent of the land has vanished into the gulf in recent decades,
and its residents—most of them Native American—were slated
to move forty miles inland.[41] The move was widely hailed as
the U.S. federal government's first effort to relocate an entire
community because of climate change.

Coordinated movements like this tend to involve relatively
small communities, since it is incredibly difficult to get large
groups of people to agree on anything, much less where they
are all going to live. A.R. Siders, a leading scholar on managed
retreat, described the coordination sessions as the most extreme city council meeting ever, full of debates about prosaic
details that are as tiring as they are important. "Anyone who
has ever served on a committee or worked with a big group of
people, it's trying to make everyone happy at the same time,"
she told me with a sigh, "and that's just a gigantic challenge.
Imagine if you're building a new town, which you might be if
you're relocating everything; are you just replicating the old
town? Literally, does the post office go next to the grocery
store, goes next to the library? Or do you reverse the order?

Oh, and people will have opinions. They will have thoughts. And they will tell you those thoughts. And they will not agree on those thoughts." We spend our entire lives making our homes just so, finding the right school district and the most efficient work commute, arranging the toaster and the blender and the coffee maker, deciding which plants to put in the front yard. Imagine doing it with every single one of your neighbors in the span of just a few months.

It can be an incredibly expensive slog. For decades, local leaders have been trying to pick up all the people on Papua New Guinea's Carteret Islands and move them fifty miles southwest, to the country's Bougainville Island. At the beginning of the millennium, watchers thought the Carterets might be underwater by 2015, but when that year arrived, a record 1,200 people were still living on the five populated islands forming a broken circle. They had been able to remain thanks to the effectiveness of seawalls, mangroves, and other protections.[42] Still, they cannot remain indefinitely. Ocean water is seeping into the ground, killing plants and poisoning the water. Storm surge is wiping out homes and vegetable gardens that grow the little local produce available. Erosion is gnawing away at the beach. Since the 1960s, younger islanders have moved to Bougainville Island in search of better jobs and better education, and the government has offered some support for Carteret Islanders to move since the 1980s. But the initial plan was slipshod and many people were unable to acquire a clear land title, so they often returned.[43] In 2006, elders led by a woman named Ursula Rakova renewed the effort with a non-profit group called Tulele Peisa, which translates to "sailing the waves on our own" in the Halia language. The group has secured enough land from the Catholic Church to relocate a few dozen families to Bougainville, putting about one hundred of them in metal-roofed, bamboo-sided houses. But it's been slow going. As of this writing, roughly two decades after the project began in earnest, hundreds are still in the Carterets.[44] Moving just 350 families would likely cost more than $3 million, a prohibitively large amount for poor people in a poor part of a poor country.[45]

Importantly, all of these movements are within the same country. There has never been a case of environment-driven planned relocation across international borders, such as might happen if Kiribati invokes its backup plan of moving to Fiji. If something like that ever happened, the details would be much more complex and infinitely more meaningful. Would children go to i-Kiribati schools or Fijian schools? Would their teachers have to pass i-Kiribati licensing exams or Fijian ones? Which flag would students say the pledge of allegiance to every morning? For that matter, would i-Kiribati people need visas just to live in their own homes? "Those really matter if the point of your relocation is to preserve your culture or preserve your tradition or preserve your connections," said Siders. "Then there's this trade-off between needing to do it right, and so wanting everyone to be engaged, and at the same time kind of needing to do it fast." The clock, after all, is ticking.

There are ways to do a planned relocation subtly, in a piecemeal fashion. That is more or less what the Federal Emergency Management Agency (FEMA) does when it buys out properties in flood-prone areas of the United States. The government is encouraging people to leave, even if not quite rebuilding the post office, grocery store, and library all in a row. The idea is simple: To reduce the prospect of future flood damage, the government can buy out at-risk houses and knock them down. These buyouts are often controversial and, like in other examples, residents often complain they don't get enough money for their property. Research also has shown that wealthier, less dense communities are most likely to be targeted for buyouts, while lower-income communities get cast aside.[46] Still, it is a top-down way of reshuffling where people live. And it can have a snowball effect; as more and more neighbors take buyouts and towns empty, the idea of moving becomes more attractive.[47]

For big cities, planners' best hopes of confronting new and changing climate threats may be some combination of small-bore policies like these buyouts combined with zoning and other rules for new buildings. Seawalls and dikes are all well and good, but even places like New York City cannot insulate themselves completely. After Hurricane Sandy hit in 2012, New York State spent $240 million to buy out and destroy 610 properties, mostly in hard-hit parts of Staten Island.[48] New zoning and insurance policies can make it more difficult to build in flood-prone areas (or those at risk of wildfires, or erosion, or whatever hazard local communities face), which impacts where future housing gets built. "Many towns grow and shrink amoeba-like; they expand in one direction, they contract in another. Well, if you expand in one direction and contract in another, and you do that enough times, you've relocated," Siders, the managed retreat scholar, told me. New York, she noted, is quite literally sinking, in some places by more than two millimeters per year, and facing a future of more storms and higher seas, which means more flooding.[49] "The idea of New York City picking up and moving is insane.

That's not going to happen," she said. "But when you think about how much New York City has grown and changed and how its footprint has changed in the last two hundred years . . . is it that crazy to think that New York might invest more in land that was northern and inland and upland, and shrink its footprint [in other directions]?" Gradually, over the next two centuries, why couldn't the densest parts of the city slowly migrate to the Bronx, rather than Manhattan? It would be an organic migration, but it could be encouraged with a little government push.

Institutions can also be rebuilt elsewhere. Belize, a small and sparsely populated country squeezed between Mexico, Guatemala, and the Caribbean, has a long coastline dead center of the Central American land bridge. The country forms the vertical axis of a right angle facing the Caribbean, adjacent to the horizontal coastline of Guatemala and Honduras to the south. As a result, Belize is nearly constantly at risk of hurricanes, which barrel through the Caribbean and slam into land. When Hurricane Hattie came rushing through the capital, Belize City, as a Category 4 storm in 1961, wind speeds were above 150 miles per hour and storm surge flooded high over the land. Buildings were buried in mud up to the third floor and as much as three-quarters of the city was flattened or severely damaged.[50] "Piles of rubble are all that remains where once there stood a city," a British newsreel intoned as cameras panned over the splinters of wooden buildings.[51]

To prevent the destruction from recurring, officials decided to rebuild from scratch somewhere else—at least partly. Belize did not become independent until 1981, and at the time the country was still a British colony known as British Honduras. To protect colonial offices and also soothe a long-festering territorial dispute with neighboring Guatemala and its U.S. backers, colonial administrators began looking for a new capital city in 1962, and soon selected a site fifty miles inland.[52] Belize City was not relocated root and branch to form the new capital, Belmopan. To this day, Belize City is a thriving and bustling city that continues to be the country's economic center, with a population more than three times Belmopan's. But it is no longer the center of government, and many residents followed when the institutions moved.

It wasn't a brand-new idea. Governments elsewhere have relocated capitals for a range of usually political purposes, often to be more geographically central, soothe regional tensions, and separate themselves from a country's cultural and economic hub. Washington, DC, represents one such example, created to host the U.S. government after early temporary stints in New York and Philadelphia. Another is Brasília, inaugurated in 1960 after rapid construction in order to move Brazil's capital out of Rio de Janeiro. Or we might look to Naypyidaw in Myanmar, or Dodoma in Tanzania, or Islamabad in Pakistan, all of which were designated to become their countries' new capitals in just the last few decades. The cases are numerous. What makes Belmopan unique is that it is likely the first modern capital to be moved primarily because of environmental vulnerability.[53]

It won't be the last. Indonesia is in the process of moving its administrative capital from the overcrowded and flood-prone Jakarta to a largely uninhabited site in the rain forest called Nusantara, on the island of Borneo, about a thousand miles across the Java Sea. Jakarta, a coastal megacity hosting more than 30 million people, is sinking fast, due to sea-level rise as well as poor groundwater management.[54] Forty percent of the city is below sea level, and some areas are sinking by as much as a foot per year.[55] By the middle of the century, as much as one-third of Jakarta could be underwater. Moving the capital will take until the 2040s and will cost tens of billions of dollars.

The biggest problem with relocating entire communities is that, time and again, people don't want to leave. No matter how threatening the encroaching seas may seem, no matter how bad the floods and storms get, people almost always have deep connections to their homes and want to stay. For Indigenous communities in the Pacific and elsewhere, relocation is a tacit acknowledgment that your ancestral homeland will soon no longer exist. "Displacements of populations and destruction of cultural language and tradition is equivalent in our minds to genocide," Tony de Brum, a former foreign minister of the Marshall Islands, explained at a 2015 international forum on climate migration.[56] Years later, as part of its efforts to design a national climate adaptation plan, the government of the Marshall Islands interviewed a significant number of island residents; only 1 percent said they wanted to leave their homes. Often, Pacific Islanders have framed relocation as the move of last resort. In-place (known as in situ) adaptations should be the main focus, with retreat only if all else fails.

The same holds true elsewhere and across time. Despite the overwhelming vote to leave Niobrara, Nebraska, in the 1970s, many residents nonetheless rankled at being pushed out. "Had we not been forced to relocate, caused by errors made at the hands of the Corps of Engineers many years ago, I could, of course, continue to do business out of my present location for many years," said one grocery store owner who estimated that, even after the buyout, he would need to borrow $140,000 to reopen his store in a new location.[57]

Moreover, there are also more nefarious reasons to clear out a neighborhood—or even a whole country. Climate change can be an easy pretense to force people from their land. As part of its plan to transform into an environmentally sustainable city, Rwanda's hillside capital, Kigali, has cleared out informal settlements at risk of landslides and flooding. Residents say they were not given compensation but are instead being pushed out of the city to make way for glimmering modern buildings and expansive apartments.[58] And while not invoking environmental threats, Donald Trump gestured at a similar argument in calling for forcibly clearing the approximately 2 million people out of Gaza to create "the Riviera of the Middle East." "It's been an unlucky place for a long time. Being in its presence just has not been good," he said alongside Benjamin Netanyahu in a 2025 White House press conference. Gazans, he added, "are living in hell. And those people will now be able to live in peace."

Even if governments truly do intend to move people for their own safety, no policy is made in a vacuum. The difference between one community getting a billion-dollar barrier system and another being told to relocate is a matter of political pressure, government priorities, and power. "Time and time and time again, there are other motives," said Erica Bower, who has researched planned relocation cases around the world. "There are political, demographic, economic, social factors that often motivate these moves both for the relocating people themselves and the facilitating actors."

Realistically, managed retreat takes time and effort and lots of money, all of which are in short supply, particularly for the most vulnerable communities. It's a solution that works best for small populations in very particular areas, where a well-meaning government or other body can step in to guide the process. It won't work for most of us. Whether in the Pacific or elsewhere, most countries are not planning to pick up everyone and drop them someplace else. The corporate world can have a say, such as when insurance providers raise rates or stop providing coverage in climate-vulnerable places. But the invisible hand of the market has proven itself woefully impotent to tackle global climate change so far, and making life more expensive is not a sustainable strategy to get people to move in an orderly way.

3

After the Flood

When it is not orderly, climate migration can look like a crowded maze of homes and shops stacked on top of one another and leaning over alleyways only a few feet wide. The busy corridors become packed with children playing and scruffy dogs looking for a place to rest. This is Korail, the largest informal settlement in Dhaka, Bangladesh's bustling and fast-growing capital of 25 million people, which has become a magnet for migrants from across the country. Approximately 2,000 people move from Bangladesh's countryside to Dhaka every day, often fleeing a mounting combination of high seas, storms, droughts, and other threats. Many end up here, in Korail, where 200,000 people live right on top of one another on just a hundred acres of land.[1]

Bangladesh is one of the planet's most climate-vulnerable countries. Large portions of its landmass seem to perch right on top of the water. The country sits at the confluence of three rivers coming down from the "water tower of Asia" in the Himalayas—the Brahmaputra, Ganges, and Meghna—as they flow into the Bay of Bengal and make up the largest delta on earth. The rivers are massive, swelling beasts that can be six miles wide and, when the waters are high, flow as fast as 180,000 meters per second, a devastating fury that can wash away buildings and farms that otherwise depend on the water for life.[2] Warming temperatures make the rivers heavier, as more snowmelt and upstream rainfall increase the amount of water making the long trek to the sea. As the rivers churn, they wash the earth downstream, trafficking silt that constantly shifts the rivers' boundaries. Every year, as much as 1,800 miles of riverbank experience significant erosion.[3] The delta has also been pounded by more periods of extreme rain, while changing patterns of seasonal monsoons have warmed the nearby Bay of Bengal and caused the sea to rise about 50 percent faster than global averages.[4] This squeezes the delta from both ends, increasing the odds of flooding and magnifying the impact of seasonal cyclones, which are growing stronger because of climate change. As storms move over the warmer ocean, they draw more energy and grow in intensity. Four-fifths of Bangladesh is less than seventeen feet above sea level, leaving vast swaths of land exposed to inundation on a regular basis.[5]

The country is also exceptionally crowded, with about 175 million people living in an area the size of Iowa. Not counting small island countries, city-states like Monaco, or similarly tiny nations like Bahrain, Bangladesh is the most densely populated country on the planet. About 3,300 people are crammed into every square mile, compared to 1,200 people per square mile in neighboring India, less than 725 in the United Kingdom, 93 in the United States, and just 10 in Canada.[6] Bangladesh is also relatively poor. While the country has made major gains in reducing poverty in recent years, more than 21 million people live on less than $2.15 per day.[7]

For these reasons, climate change is especially pernicious. People living cheek by jowl in poorly built houses are prime candidates to be hit hard by disasters. Many people in rural areas also depend on agriculture, so creeping sea-level rise that replaces freshwater stocks with seawater can undermine their livelihoods and rob them of their wealth. In rural areas in particular, the signs of a changing climate are everywhere. Ninety-four percent of villagers say they have experienced floods in the last three years, and 91 percent experienced riverbank erosion. Additionally, nearly four-fifths say that rainfall has increased and the weather has gotten hotter.[8]

People are being forced out of their homes. In northwest Bangladesh, for instance, the average household has been displaced by river erosion 4.6 times.[9] In just 2019, more than 4 million Bangladeshis were displaced by climate-induced disasters.[10] About 10 million people in Bangladesh were considered to be climate migrants as of 2022, and there may be 19.9 million internal climate migrants in Bangladesh by 2050, according to the World Bank, at which point more people would be migrating for climate reasons than any other factor.[11] Bangladesh has been stricken by a mix of both slow-moving environmental challenges and rapid-onset disasters like cyclones. Combined, they can destroy people's homes and livelihoods and drive them elsewhere—not in a coordinated process of managed retreat but in a panicked rush.

Bangladeshis have migrated for decades, particularly on a seasonal basis, and the vast majority have stayed within the country. While people with an education and some connections might hope to move abroad—particularly to Saudi Arabia and elsewhere in the Middle East, where millions of Bangladeshis and other South Asians work in construction, domestic service, hospitality, and other sectors—the vast majority go to other parts of Bangladesh, especially the cities. In fact, one of the defining global trends of the last few centuries is for people to move into urban centers and away from rural areas. Since about 2008, for the first time in history, most of the world has been living in urban areas.[12] Climate change combined with newly advanced economies is making this trend more acute in countries around the planet, as farming is simply becoming outmoded. From 1950 to 2018, the number of people living in urban areas worldwide rose from around 751 million to 4.2 billion.[13] Cities are projected to add another 2.5 billion people by 2050.[14] Not all of these people are migrants, of course—many are the children of current city dwellers—but it's clear that the future is urban.

Dhaka's government cannot keep pace with the growth. There is a local saying that "in Dhaka City, money flies," according to Mahmudol Hasan Rocky, a researcher with the Dhaka think tank Refugee and Migratory Movements Research Unit. The city is a shining symbol of Bangladesh's desire to leapfrog out of poverty and also of the dangers inherent in such a jump. There is a growing middle class, thriving cultural scene, and signs of a city on the upswing. But there is also deep poverty, visible in the elderly women knocking on car windows to beg for change and in the crowded, sprawling slums. Korail is perhaps a perfect encapsulation of the inequality that often comes hand in hand with rapid urbanization. It lies across the thin Banani Lake from a wealthy neighborhood with ten-story apartment complexes made of smooth glass and steel. When the sun sets in the evening, the glittering lights on well-heeled roof decks sparkle in the twilight as they loom over the flimsy shacks across the water.

Informal settlements like Korail are a modern-day version of the Little Oklahomas that populated Dust Bowl–era California and mirror the crowded neighborhoods of Kiribati's cramped Tarawa atoll. Similar ramshackle neighborhoods are likely to emerge whenever a rush of people move to an area without land or cash or connections. The conditions in Korail can be grim. At least one-third of children are not in school, and most don't reach the high school level. Most people earn the equivalent of about $120 per month.[15] Trash piles in corners of alleyways and dozens of people share a single toilet.

Many migrants in Korail are here only temporarily. Some come just for a few months at a time, when work runs dry in their rural homes. Others intend to stay for good but cannot find their footing in the city and move back home when things don't work out. Or they will just dip their toe in before deciding to live there full-time. A typical pattern is for a young man to move first and then bring along the rest of his family once he's established himself. It's a step-by-step process that takes many forms and often moves at an uneven pace. Historically, many migrants in Dhaka have gone into the garment sector, stitching together clothes that are sold around the world and which may be sitting in your closet right now. Construction is also a big lure, creating a novel feedback loop: more migrants building new buildings for more migrants to live in.

But Korail is also a place of human beings, a place of hope and joy. A seemingly endless number of children play in the narrow alleyways, spinning tops and throwing cricket balls and filling the streets with laughter. There are barbershops, clothing stalls, mosques, and men gathering to smoke and chat and sip tea. People share gossip and jokes and complain about politics, and because all of them live on top of one another, they also laugh and yell and fret together too. When the call to prayer rings out, it seems to come from every direction,

echoing chaotically and beautifully through the corridors like
a John Cage piece come to life.

It was in Korail that I met Tonni, an extroverted and asser-
tive woman with a take-charge attitude. She came to Dhaka as
a child three decades ago, when the Louhajang River burst its
banks and washed away her family home in Tangail District,
sixty-odd miles away. Tonni's grandfather, who still lives in
the ancestral home, has been displaced four times. Now Tonni
works at a local NGO and her husband has a good job at a
shipping company. She has three kids and the eldest boy is preparing to graduate high school. Next up for him is either the
military or, Tonni hoped, a chance to study in Europe, ideally
Germany or Italy. It would cost $30,000 to make that happen,
she reckoned, a seemingly astronomical sum, but somehow I
got the sense that hers was not merely a pipe dream. Tonni's
family also receives some income from land she's built up and
is renting out, which is helping her save for the prospect of her
son going abroad. When I asked what her life might have been
like if she had stayed in Tangail, she was unequivocal; in her
village, there would have been nothing but floods and poverty.
Now she has a solid job, her kids are getting a good education,
and they are preparing to clamber up the socioeconomic lad-
der. Squint a little, and her story is not all that different from
that of the Okies who eked out a living in California so their
children and grandchildren might move into the middle class.

There are many people with stories like Tonni's in Korail.
After meeting her outside a tea stall where people were gathering to watch a cricket game on TV, I spent nearly two hours
with her and several friends who told me how they had similarly been forced out of their rural homes by floods and storms
and moved with their families to Dhaka as an escape. We sat

in the one-room home of Parven, a forty-something woman with a black headscarf and fading henna on her hands. She came to Dhaka as a child, in 1988, after her family house in the southern Barguna District was swallowed in one of the worst combinations of environmental disasters to strike Bangladesh at the time. Catastrophic flooding due to heavy rains was followed by a major cyclone that tore through the south of the country with wind speeds nearing one hundred miles per hour. Thousands of people were killed, hundreds of thousands of acres of farmland were destroyed, and millions of people were left homeless.[16] Everything was wiped away, Parven told me, so she moved to the capital along with her parents and six siblings.

Her new home is small, just a simple room with tin walls and bamboo beams for support. There is room enough only for a bed, dresser, some shelves, and a few pots and pans. There is a cook stove outside, where she surreptitiously and generously made me tea. A window looked out onto the communal water pump, where neighbors' comings and goings were clearly audible. Another friend, slightly older, had been married at thirteen and had no other options when their family home outside Dhaka washed away. Another came in 1998, after Bangladesh's most devastating flood of the twentieth century, in which one thousand people were killed, 1 million homes were damaged, and 30 million people struggled to rebuild.[17] Water wiped out her house and the rice paddies where her father worked in the Bhola District, also in the south, she said.

Still, Dhaka is not always the sanctuary it might promise. Traffic clogs the capital's streets, as cars and tuk-tuks use horns in place of turn signals, creating a constant clatter that sounds alternatingly like a cacophony and a symphony. In parts of

Dhaka, ambient air quality is as bad as smoking two cigarettes per day.[18] The city is hard at work building an expansive metro rail system and new highways to relieve the strain from the traffic, but it is unclear whether the new construction can keep up with how quickly the city is growing. Then there is the heat. Even in a sweltering country like Bangladesh, Dhaka stands out for being especially heat-prone. This is partly a result of the urban heat-island effect, in which reflective surfaces, black asphalt, car exhaust, and humming machinery keep and trap heat. In 2023, temperatures hit a humid 105°F, the highest in six decades.[19] Few people have air-conditioning, and especially not in slums like Korail, where tin walls make people feel as if they are trapped inside an oven. The capital is also expensive for people coming from poor areas with little money in their pockets. Still, many residents told me it would be nearly impossible to make as much money anywhere else. And so here they sit, breathing in smog-ridden air and baking in the heat and paying a lot of money to do it.

Floods reach the capital, too. The city's topography makes it especially prone to being waterlogged, and informal settlements near the lakes and along the Buriganga River that runs just south of town are particularly vulnerable. As much as three-quarters of the city went underwater during annual monsoons in previous years.[20] Tonni's friend Parven was in her twenties when the devastating 1998 flood hit. She and her family were in Dhaka then, where half the capital was submerged in waters that lingered for weeks, creating a pervasive and foul mix of sewage, garbage, and mud. Relief officials at the time spoke of an "unimaginable squalor of mud and filth where living a normal life is worse than a nightmare."[21] Parven remembered water reaching up to her head. She spent weeks on a bed raised by bricks to stay above the high waters, then months in a nearby elementary school until the floodwaters subsided.

Big cities like Dhaka are one option for people fleeing disaster, but many will go only a couple of miles away from home. Such was the case of Monjurul, a fish merchant with the good looks of a classic movie star whose house in a rural part of Satkhira District along Bangladesh's southwest border with India was swept into the river by floods seven years ago. I met him in the courtyard of his new home in the village of Akashlina, about twenty miles away from his family's land. To stay and rebuild the old house, he would have needed to find a new line of work, he told me in a quiet but firm voice, but he had never done anything except buy and sell fish. After the flood, he rented a home in a small community near the Indian border, but the monthly payments were difficult for a man whose income fluctuated with how well the fish were biting. So he moved to Akashlina. He paid about $3,500 for his current place, taking out a range of loans in his name and those of his family members, as well as relying on informal lenders. The move put the family in a financial tailspin, but they were slowly recovering. His eldest son, eighteen, was still in school but had fallen behind by two years due to the displacement.

Monjurul's new home is a nice one, built of thick concrete with a breezy courtyard and small potted plants scattered about. When we spoke, a duck pecked at the weeds nearby and a puppy yipped in the background. I asked Monjurul if he considered himself a climate refugee, which is the term that my translators and I had been using to describe his situation. But this notion did not translate. "No," chuckled my translator and fixer, Allen. "He's trying to make money and buy land because he has sons." The precise cause of Monjurul's predicament was immaterial; what mattered to him was finding a way to earn a living and leave a legacy for his children.

Unfortunately for Monjurul and his family, they live at the edge of the Sundarbans mangrove forest. Waters from Bangladesh's major rivers meet here, in a nearly four-thousand-square-mile area of interconnecting waterways and trees and delta on Bangladesh's border, and then crash into the sea coming up from the Bay of Bengal, creating tidal rivers that flow back and forth several times per day, alternately bringing freshwater, then salt water across the mangroves and mud flats. Around 7 million people live in or around the Sundarbans, working small farmlands, fishing businesses, or a range of other jobs.

Water is everywhere here. Even one hundred miles from the coast, the roads are flanked by wide rice paddies sprouting from ankle-deep pools. Giant ponds that locals use to wash and bathe collect thin layers of algae in the shade. Rivers cross the land like railroad tracks. Often, the only thing holding back the water is a small mud retaining wall a couple of inches thick. The wide, sweeping embankments are the only features making any effort at altitude. Fat, heavy banana leaves droop lazily in the near-constant haze.

For the millions of people living around the Sundarbans, the forest is "like a mother," explained Abdullah Harun Chowdhury, an environmental scientist at Khulna University on the forest's edge, a few hours' drive away from Monjurul's home. Historically, the waters have been diverted to feed the rice paddies that provide critical food. Many locals also depend on the forest for fishing, catching crabs, gathering honey, collecting

wood and palm fronds, or other purposes. Over tea and bis-
cuits in Chowdhury's cramped office, where stacks of yellowed
papers competed for space with bottles of sediment samples
and dried coral, he relayed the cocktail of slow- and fast-onset
environmental crises affecting southwest Bangladesh. The
slow crisis is the increasing salinity of the groundwater, he
said, as rising seas push salt water deeper into the land and
goad the high tides farther and farther up the delta's mouth.
Across the country, soil salinity has increased 26 percent since
the early 1980s.[22]

Even with all the water around, none of it is safe to drink,
pushing people (women and girls, mostly) to walk for hours to
functioning wells. Nationwide, about 80 percent of Bangladesh
villagers say they have problems obtaining drinking water.[23]
The high-saline water has caused a slew of health problems
including skin issues such as scabies and fungal infections as
well as stomach trouble. Many households collect rainwater
in giant drums distributed by aid groups, sometimes in bright
pink and orange that offer a surprising neon accent to rural
tin shacks. The evermore brackish water has also killed the
rice paddies that many locals have depended on, pushing them
into farming shrimp and crabs instead. But crabs and shrimp
are cash crops that can't always ensure farmers have enough to
eat. It also takes fewer people to tend to shrimp farms, meaning there are fewer jobs to go around. And this process speeds
up the overall soil salinity, allowing larger quantities of salt
water to pervade the ground and bleed into neighboring crop
beds, killing more rice paddies. Matters were made worse by a
flood-defense system involving high embankments surround-
ing low-lying tracts of land, called polders, which was im-
ported by Dutch advocates beginning in the 1960s. While the

system largely succeeded at preventing major floods, designers failed to account for the natural tidal flow of the rivers. Silt that would otherwise have been deposited on the floodplains instead sunk to the bottom of the riverbed where it clogged up drainage coming off the fields. The consequence is that, whenever storm surge has pushed salty water over the embankments and as more fields have been converted into shrimp farms, the high-saline water has often been trapped in the fields and prevented from flowing out to sea. To compensate, some communities and organizations have begun intentionally flooding the polders on a regular and managed basis, to spread out the silt and restore the rivers' natural shapes.

And then there are the cyclones, which descend on the mangroves every few years to wreak havoc. In the last twenty years, the area was walloped by major cyclones Sidr, Aila, and Amphan, plus multiple smaller events. Each individual storm is terrible enough on its own, flattening houses, killing livestock, and smashing embankments. Across the region, the government and aid groups have built brutalist two- and three-story cyclone-resistant shelters that loom over the marshes and can withstand the otherwise devastating 140-mile-per-hour winds. As a result of these shelters and better early warning systems, the number of people dying in cyclones has dropped dramatically over the last decade or two. But people can't bring their homes to a cyclone shelter. And the combined weight of cyclone after cyclone compounds to knock people down before they have a real chance to get back on their feet. For many inland communities, the forest itself acts as a "green wall" against the wind, Chowdhury said, but the accelerating tidal surge can pack a wallop all its own, leaving homes far inland flooded with feet of stormwater. "During the cyclones, basically high tides are coming, some areas become contaminated by the saline water permanently. Waterlogged conditions are happening," he said. It's simply too much water, of the wrong kind, coming in the wrong way, and staying for too long.

If the forest provides some protection, it also carries its own dangers. Globally, if the Sundarbans are known at all it is likely for the population of around one hundred Bengal tigers that prowl its mangroves and serve as something of a national symbol for Bangladesh. But these animals are coming into closer contact with humans as people trek deeper into the woods, and as the tigers' natural habitat gradually erodes.[24] Virtually everyone I met who goes into the forest had a story for me about a tiger attack. Older people also have stories involving roaming gangs of pirates, who used to menace the Sundarbans but have generally disappeared since the government launched an amnesty program in 2016.

Facing these kinds of conditions, people "are bound to migrate," Chowdhury said. "If they do not go, maybe sometimes their lives may be at risk. Their family members are at risk." By 2050, at least 17 percent of residents of Bangladesh's coastal areas are predicted to be displaced, adding up to millions of people.[25]

People who migrate to Dhaka or farther inland are able to leave their waterlogged villages because they have the social connections, money, and other resources to do so. Through various loans, Monjurul was able to cobble together enough money to afford his current property. When we spoke, he had been hosting his brother-in-law, whose home had been destroyed in a cyclone but who was lucky enough to have a family member to help him get back on his feet. The residents of Korail I met similarly were able to move because they had money or some help.

Not everyone is so lucky. Moving out of harm's way would be impossible for a woman named Nur whom I met in the Sundarbans. She has moved a total of nine times in response to floods, but each move was just a couple of hundred feet inland, rebuilding as she could on whatever bare space of earth became available. We met outside her roughhewn house at the top of an embankment overlooking the Arpangasia River, one of two waterways that squeeze her small community of Pratapnagar, a poor, cyclone-plagued area I traversed on the back of a motorbike one warm and hazy day in February. For years, life here has seemed unsustainable. Many men leave for up to six months at a time to work in far-off brick factories or rice paddies. Those with a bit more education leave and often never come back, joining the long chain of previously rural residents who now call cities home. Moving is not possible for someone like Nur, with virtually no money or connections outside her village. She has always lived just a few feet from the river's edge, but each of her previous homes was swallowed whole when the waterway expanded. For two years she, her husband, and their three children lived out of a boat they had dragged onto the shore. Nur has an infectious exuberance and incredible charm that easily transcended the language barrier between us. We met at the end of the day, as the sun descended behind the horizon, and she was quick to offer me a chair—perhaps the only one the family owned—while she stood and spoke to me. It was a white plastic Monobloc patio chair, the kind of thing that costs a few dollars at a Western supermarket, and she had restitched together the arm and back where the brittle plastic snapped in two. Behind her a handful of small, thumb-sized fish and half-inch shrimps dried in the setting sun.

She is thirty. Her eldest son is eighteen. Do the math and you'll realize that she gave birth at an age when most girls are just beginning middle school. Early marriages are a major problem for girls in this area, and a silent tragedy of climate change in general. About half of girls in Bangladesh are married by the time they hit eighteen, and child marriages in the vulnerable coastal regions spike 39 percent after climate-induced disasters.[26] For the poorest of the poor, marrying off children can be a way to help offset the economic burdens of losing your house, crops, or other assets. It can mean one less mouth to feed when times are tough. In cultures where bride prices are common, marrying off a daughter can also lead to a quick injection of cash when parents need it most. There also tends to be an uptick in sexual violence after major disasters, especially in crowded evacuation shelters with little privacy and lots of desperation; some families may consider child marriage to be a way to protect both their daughter and the family's reputation.[27]

Nur's house now sits twenty feet above the river, perched on a wall of sandbags next to a pond breeding the small fish that she dries and sells. A much bigger sandbag wall, towering five stories into the sky, sits a few miles away, on an embankment the government is slowly trying to rebuild. The barrier burst during Cyclone Amphan in 2020, flooding several houses and inundating paddy fields with poisonous salt water. One of the destroyed houses belonged to Abdul, a reedy sixty-something with tired eyes. The land had been in his family since the times of his father and his grandfather, he told me, but in the span of two days everything was lost. Now he and his wife live a quarter mile away, in a shack made of sticks and tarpaulin high up on the new embankment. The flooding wiped out all the trees in his area, so they are totally exposed to the baking sun. His new house sits precariously over the river, fifty feet above the water at low tide. It would be a beautiful place to live if it weren't on the edge of disaster.

When I met Abdul he was sifting through a tub of water to collect baby shrimp and separate them from the river scum. He and his wife collected the shrimp from the river at five o'clock that morning, as part of work that typically earns them about two dollars per day—or half as much when times are hard. His two sons realize there is little future for them in Pratapnagar. They have both left, telling Abdul that their home village is too dangerous to live in. They invited him to come with them, but he didn't know what he would do for work any place else.

Thousands of villagers have left for good. A local community leader said at least five hundred families migrated away after Amphan, believing their future here had been washed away along with the current. At high tide, the water continued to reach waist height for two years after the cyclone, forcing people to put bricks underneath their beds to stay above the waterline, and making them dependent on boats to get around, as if they were in a South Asian Venice. But those boat rides weren't free, and even the 50¢ fee that canoe paddlers charged added up over time. Meanwhile the stocks of fish and shrimp—not to mention the rice paddies—rotted. Even in a place as vulnerable as Bangladesh, Pratapnagar stood out. "Pratapnagar's climate woes never end," a local newspaper moaned in 2023.[28] Enrollment at the local elementary school dropped by half. Prospects for the future have seemed dim.

*   *   *

One of the cruelest things about water is the way it can simply erase someone's land off the face of the earth. In places like Bangladesh, where so many people are struggling to live in a constrained area, the unequal division of land ownership is directly tied to who stays and who goes. Because more than one-third of the country depends on agriculture to earn a living—compared to just 2 or 3 percent in North America and Western Europe—owning land has been an essential way to ensure that a rural family could put food on the table; not owning land virtually ensures they stay trapped in poverty.[29] Across the country, about one-tenth of the population—more than 4 million households—are landless, and tens of millions more are sharecroppers who either pay other people to farm their land or work just a tiny plot about half the size of a basketball court.[30] Landlessness in some places has been increasing in part because of escalating climate disasters, as well as the country's booming population. And when landlessness increases, so too does migration.[31]

Questions about who owns land and which plots in particular are often deeply contentious in developing countries, where official recordkeeping may be poor and where recent conflicts mean people took and fled property relatively recently, jumbling the claims. In Bangladesh, which achieved independence from Pakistan in 1971 and was previously part of the British Raj, significant fault lies with colonial decisions about who was able to acquire wealth and who was prevented from doing so.

This was the narrative impressed upon me by Zahid Shashoto, a passionate young development worker with an Oxford-educated lilt. "The reason why a cyclone is pushing people towards migration, the reason why a tidal surge or a breach of embankment or anything of that sort is actually forcing people to migrate is the structural marginalization of people," he said. "Poverty here is structural." When people do own land, it tends to be small tracts. In rural areas, 89 percent of landowners own less than two and a half acres—roughly two football fields, scarcely enough to grow a little rice.[32] That makes it hard to rotate crops, invest in climate-resilient infrastructure, and otherwise prepare for the worsening climate. Many landless people depend on the complicated Bangladeshi system of *khas* land which can be leased from the government and was created to assist people with few assets. While the *khas* land system can be an important tool to lift rural families out of poverty, the opportunities for corruption and slow-moving bureaucracy mean its promises are not always realized.

Moving likely won't change that underlying marginalization, said Shashoto. Especially when they move to cities, people are hardly likely to gain any new rights. They might be able to earn higher incomes, get a better education, and be less vulnerable to disasters, he said. "But they do not have any legal rights to those lands [in informal settlements] and they can be evicted any time government wants, or local elite wants. They do not have access to social safety net services," because even if they might be eligible for various benefits they often have to formally reregister their residence with the government, which can be difficult, or may not understand the process for applying. Moreover, slums may not offer legal access to utilities like gas and electricity. "What I'm saying," he continued, "is these are people who don't have an identity in the city, don't have the dignity that they want to have. They just have better income and are less vulnerable. But what about all these other factors? And how can you force someone to migrate or encourage someone to migrate without ensuring that he or she will have rights to their land and dignity?"

In Bangladesh I spent a day on Gabura, a roughly ten-square-mile island a few miles downstream of Pratapnagar. Gabura too is regularly ravaged by cyclones and riverbank erosion, and the soil's rising saline levels are a constant challenge for locals. In 2009, Cyclone Aila killed 28 people on the island and made 25,000 homeless. Fifteen years later, many villagers still talked about their homes and farms that had been destroyed by the storm. Shortly after it struck, the chief executive of Oxfam Great Britain at the time, Barbara Stocking, described Gabura as "a terrifying vision of a world devastated by climate change." "In the small, impoverished community of Gabura in Bangladesh, the concept of global warming, often only words on a screen or in a newspaper to us, is an all too bleak reality."[33] For a moment, Gabura seemed to be the epicenter of Bangladesh's climate disaster. As in Pratapnagar, the riverbanks were eroding and dragging homes into the sea. The island was literally collapsing, losing about two-fifths of its mass to the riverbed.[34] The groundwater was too salty to sustain the rice paddies and the jungle was growing more dangerous. Two-thirds of men on the island in 2023 earned less than $45 per month.[35] In 2023, scientists predicted that, without assistance, the entire island could be subsumed by water in the next two years.[36] While exploring the island I met Gulum, a member of the local community council, who told me that one out of every twenty or thirty families had left because of the island's climate threats. "They have already gone," he said. Two types of people migrate away, he said: "high-risk families" and "landless families" whose land is "gone already into the river."

But that was the past, and Gulum was looking to a future that had miraculously become more hopeful. We spoke on top of a new embankment that the government had been racing to construct, as part of a massive $400 million project across southwest Bangladesh. The effort has upgraded and replaced more than 250 miles of seawall, benefiting about three-quarters of a million people.[37] In Gabura, a new seawall will encircle the island and new canals will relieve extreme tidal surges, to stop and redirect the river's racing and rising waters.[38] Locals refer to the construction as simply the "megaproject." Construction should finish in 2026, locals hoped. It may not be the MOSE system protecting Venice or New York City's Big U, but the project could pay major dividends for this remote endangered community. Everyone I talked to on the island had high hopes that the megaproject would finally offer some certainty for their future, allowing them to stay in the place where they were born and raised. Work seemed to be moving at a frenetic pace, with multiple earth movers digging out trenches and heavy concrete blocks being moved into the breach.

The hot sun beat down on us while Gulum and I chatted, both of us peering at each other from behind our dark sunglasses. A balding man in his mid-fifties with an electric-yellow button-up shirt, he was passionate about the megaproject and was eager to express how it would revolutionize his small hometown. "If [the] riverside embankment is strong, I think that people are protected." His neighbors won't have to flee anymore, he insisted, as the midday call to prayer sounded in the background, and maybe some people who left in previous years might come back.

Still, village life has its limitations. Gulum's son had long since left Gabura and was studying electrical engineering at a university in Dhaka. No matter what happens to the embankments, he would not be coming back. There are no jobs for someone with a university education on this small rural island, not to mention the comforts of air-conditioning, highways, and other modern conveniences. In Gabura, "no have road, no have good house," Gulum said with a laugh. "He don't like." But Gulum himself would not follow his son off the island. He was born here, he said, and he'd lived here all his life. "Do, die, everything in Gabura."

Not everything can be rebuilt or protected with a seawall. Some Indigenous tribal communities live in the Sundarbans, accounting for a portion of the dozens of minority Indigenous groups in Bangladesh who comprise about 1 percent of the total population.[39] The vast majority of the country is ethnically Bengali, and Indigenous people are not explicitly recognized in Bangladesh's constitution; many tend to be poor, often face discrimination, and historically have suffered from human rights abuses.[40] Climate change has accelerated the forces tearing their groups apart and further jeopardizing their traditional ways of life. "We have also lost our culture," said Krishnapada Munda, a young-faced social worker and activist with a calm but intense manner. His people, the Munda, practice an Indigenous religion and in India are considered to be very low caste. Centuries ago, Krishnapada's people came from India, and there are only about five thousand Munda people in this part of Bangladesh—a relatively tiny number, which makes their traditions and heritage all the more vulnerable to disappearing.[41]

Krishnapada spoke with me not far from his home outside the Sundarbans, in a cool, tiled office used by local journalists. Dressed in cargo pants and a black hoodie proclaiming "Time Is Life," he relayed the various ways that the changing mangroves were chipping away at his people's traditions. "We have a big festival named Karam festival. This is a big festival, it lasts about seven days. But our belief [is] that we need a tree named Karam, but this tree has died due to saline water." The tree, formally known as *Nauclea Parvifolia*, grows up to fifty feet tall, with dark green leaves and yellow globes of fragrant flowers. It has been used for medicinal purposes including as an anti-inflammatory and antidiabetic. Krishnapada's people tried replanting a tree from another district, "but that tree also has died," he said. "So now with this, we cannot celebrate this festival, Karam." Instead of a seven-day harvest festival involving dancing and singing, his people now do an abbreviated version with just one day of prayer.

The Munda people consider the forest to be a god, and its gradual decay is like a slow but steady act of deicide. Saline water has also killed off snails and other traditional foods they enjoy. And by wiping out the rice paddies, it has taken people's jobs and sources of sustenance. Because they traditionally speak their own language, worship their own religion, celebrate their own festivals, and eat their own food, Indigenous groups like the Munda often find it hard to benefit from the loans and other assistance their neighbors might use to switch from farming rice paddies to cultivating shrimp. They tend not to have assets to guarantee a loan, so they are forced into high-interest arrangements that are difficult to repay. That makes the prospect of going elsewhere all the more attractive. "There is not enough work, they cannot return back [the loans], so . . . they leave this country," Krishnapada said. "Some things are very threat-ful for us. And we need resilience power, adaptation and resilience power. . . . If we increase our resilience and adaptation, we believe that we will struggle against any kinds of cyclones, floods, or river erosions, and our next generation will get a good place."

For the Munda, poverty and marginalization have contributed to a climate-induced scattering that is further eroding their traditional life. Their challenges echo those confronting natives of Kiribati and other low-lying islands. For them and for many other people, climate migration is not just about finding a new home but also reckoning with the loss of their heritage. Communities' identities are at risk of crumbling along with their houses. When all that disappears, what is left?

4

There Is No Such Thing as a Climate Refugee

Once upon a time, international migration was easy. Until a little more than a hundred years ago, governments were relatively lax about who came across their borders, why, and for how long. Limiting migration was logistically difficult and typically considered counterproductive, preventing workers and money from traveling wherever they might be most profitable. "By the existing law of Great Britain, all foreigners have the unrestricted right of entrance and residence in this country," the secretary of state for foreign affairs, Granville Leveson-Gower, Second Earl of Granville, proclaimed in 1872.[1] There was also generally a lot less migration, in part because it was much more expensive. In any case, what was a country to do with someone at their border they didn't want? Pay a lot of money to put them on a ship going in the other direction? Some countries tried to limit immigration based along racial lines, such as the 1882 Chinese Exclusion Act in the United States. But these were largely one-off policies designed to keep out certain people, rather than systems in which everyone was excluded by default and only specific individuals were allowed in.

World War I changed things. As it became easier for individuals to move, governments started locking their doors. Increasingly, immigrants had to pass certain tests, carry passports, or otherwise obtain approval from the destination country before entering. Being allowed in became the exception, something that had to be affirmatively declared. Now immigrants tend to fall into one of a handful of discrete legal categories: labor migrants, for people moving for a job; family migrants, for those following a spouse or other family member who is already legally present; international students; "leisure migrants," who tend to be retirees or others with money and simply want to enjoy a new country's good weather, tasty food, and other treats; and humanitarian migrants, including refugees and asylum seekers. There are also temporary migrants such as tourists, business travelers, and others who receive short-term visas of just a few weeks or even days. Countries have different visas and schemes for each of these different categories and offer a dizzying array of rules, prescriptions, and requirements.

None of them are designed to accommodate people escaping climate change. In public discourse, climate migrants are often grouped with humanitarian migrants, under the thinking that their movement was forced by powers beyond their control, just like refugees fleeing war. But that is technically inaccurate. No one on earth is a refugee because of climate change. Legally speaking, it doesn't matter. The document creating the global refugee regime in the aftermath of World War II, the 1951 Refugee Convention, defines a refugee as a person who, "owing to well-founded fear of being persecuted for reasons of race, religion, nationality, membership of a particular social group, or political opinion, is outside the country of his nationality and is unable or, owing to such fear, is unwilling to avail himself of the protection of that country."[2] There is no mention of climate change, or the environment, or the sea swallowing your home.

These were not major concerns at the time. The refugee system was built in part to respond to the mass displacement caused by World War II. It fully blossomed during the Cold War, as Western countries were eager to position themselves as safe havens for people escaping the Soviet bloc. The number of refugees worldwide has fluctuated over time, peaking at the end of the Cold War, declining through the 1990s, and then shooting up precipitously since the Arab Spring and the Syrian civil war in the 2010s. Today, most refugees live in countries next to the ones they left, and three-quarters are in low- and middle-income countries. While refugees are among the most vulnerable people on earth, they are nonetheless supported by a system of aid groups, governments, and international agencies such as the UN Refugee Agency, UNHCR. Depending on their situation, refugees tend to end up with one of three fates: they return to their country of origin if the situation improves; they integrate into the local community wherever they are; or they are sent by plane and resettled in a new country such as the United States or Canada.

Climate migrants have no such framework. There is no UN agency to care for, screen, and resettle people whose homes have been destroyed by climate change. In other words, there is no "right way" for these people to flee disaster, at least not across an international border. If they want to go to another country, they must typically secure a regular visa through one of the established legal categories, such as by getting a job or joining family already abroad. For many, this is simply impossible. So if they need to move internationally, many migrants are forced to do so illegally.

To be clear, this is not a problem that most migrants face. Most countries are big enough for people to move to another neighborhood or town less exposed to environmental hazards. Generally speaking, most migration happens within a country, rather than between countries, and this tends to be the case with climate change, too. Moving within a country is easier: There usually isn't paperwork involved, people tend to speak the same language, and so on. Assuming you can afford it, you just pick up and go. But this isn't always possible in the tiny atoll nations of the Pacific, where there is literally no place to run. And even for people in bigger countries, moving internationally may be more promising for a whole host of reasons. International migration has been an appealing option for people throughout history, especially for people facing disaster.

There is an effort afoot to simply expand the legal definition of a refugee to include people displaced by climate change, so climate migrants could become eligible for international protection. But experts worry that the system is already under intense threat and can hardly bear to be more expansive. Even with our current narrow definition of a refugee, numbers have grown to a record 38 million as of 2024, according to UNHCR.[3] As it is, many countries are suspicious of refugees, and in dozens of places most people tend to agree with the notion that immigrants coming as refugees "really aren't refugees" but simply disguising themselves in order to find a job or take advantage of government benefits.[4] You don't have to look far to see the backlash. From Donald Trump to Marine Le Pen to Viktor Orbán, Western leaders are eager to reduce immigration and make it all but impossible for refugees to receive protection. "The asylum program is a scam," Trump said in his first term, referring to the decades-old legal framework for protecting humanitarian migrants who arrive at the United States' front door. "Some of the roughest people you've ever seen. People that look like they should be fighting for the UFC."[5] Trump all but shuttered U.S. refugee resettlement in his first term, ending the country's historical status as the top receiver of refugees. On day one of his second term, he suspended the program indefinitely.[6]

Even if the legal definition of a refugee were to be expanded, there is no guarantee it would provide a pathway to safety. Of the 38 million refugees in the world, fewer than 204,000 were submitted by UNHCR for resettlement in 2024, about 0.5 percent of the total.[7] Resettlement is typically considered the last resort for refugees, but it's clear that simply becoming a refugee has little bearing on whether someone will actually be resettled in a new country.

These were the kinds of tensions that New Zealand leaders were grappling with in 2017. Fresh off elections that saw the Labour Party of Jacinda Ardern scrape its way into power, unseating Bill English's center-right National Party for the first time in a decade by forming an unusual coalition with the Green Party and the anti-immigrant New Zealand First, the government had promises to keep. Ardern had campaigned on reducing net migration by 30,000—no small feat for a country where that year's gain of 70,000 immigrants was record-setting[8]—and Ardern entered the prime minister's leafy Premier House only after weeks of negotiations that led to an alliance with NZ First and saw its founding leader, Winston Peters, take office as deputy prime minister and foreign affairs minister.[9] Peters has been called the "Donald Trump of the South Pacific" for his populism and knee-jerk opposition to immigration.[10] On the other side, Labour also had debts to pay to the Green Party, which had wanted to increase New Zealand's refugee intake nearly sevenfold, from 750 annually to 5,000, calling the previous policy "a stain on our reputation as a caring country." [11]

The result was a potentially landmark announcement: A new legal program would allow Pacific Islanders displaced by rising seas to come to New Zealand. It would be "an experimental humanitarian visa category for people from the Pacific who are displaced by rising seas stemming from climate change, and it is a piece of work that we intend to do in partnership with the Pacific Islands," Green Party leader and Climate Change Minister James Shaw said just a few days after the results of the election became clear.[12] Only one hundred visas would be offered annually, at least at first, which was minuscule considering the tens of millions of people at risk globally. But it was enough for Peters and his NZ First supporters to stomach. And, if successful, the visa could have inspired others and been a model for the world. "The whole world is pulling for you," cheered Al Gore, the former U.S. vice president turned climate campaigner.[13] The small experiment could "trigger the era of 'climate change refugees,' " the *Washington Post* blared.[14] "The lives and livelihoods of many of our Pacific neighbors are already being threatened, and we need to start preparing for the inevitable influx of climate refugees," UNICEF New Zealand head Vivien Maidaborn argued at the time.[15] "There's a commonly made argument that New Zealand is too small to have an impact on global climate change. It is true to say that we are not a world leader in terms of overall emissions. But we are not too small to show leadership."

The announcement came at an auspicious time. Just days earlier, New Zealand's government ruled that two families from Tuvalu were not eligible to become refugees because they did not face persecution in Tuvalu based on the conditions spelled out in the Refugee Convention: race, religion, nationality, membership of a particular social group, or political opinion. The Immigration and Protection Tribunal reviewing the case "accepts that returning to Tuvalu will pose challenges for the appellants both socioeconomically and in terms of the negative environmental impacts of climate change," it ruled. However, "there is no basis for finding that any harm they do face as a result of the adverse impacts of climate change has any nexus whatsoever to any one of the five Convention grounds." [16] In other words: Sorry, but we can't help you.

Similar cases had played out in previous years. Some had happier endings. In 2014, New Zealand granted residence to Tuvalu native Sigeo Alesana and his family after they claimed to be threatened by climate change, in the country's first decision of its kind. But that situation was unique. Alesana and his wife had moved to New Zealand in 2007 but overstayed their visas and fell out of legal status in 2009. As such, they had strong ties to New Zealand in ways that new arrivals would not; his two young children had been born there, and his mother and sisters were also living in the country. These and other factors—not solely the fact that their island might one day soon cease to exist—helped convince the Immigration and Protection Tribunal that they deserved to stay in the country. In fact, Alesana and his family weren't granted refugee status, but allowed to stay in the country on "exceptional humanitarian grounds," a novel feature of New Zealand law.[17]

With this background, the new climate refugee visa seemed like a game changer. At long last, the government was resolving a problem that had nagged it for years, in the process positioning New Zealand as a global leader on climate change and meanwhile also keeping the number of new arrivals minimal, so as not to upset voters anxious about immigration. A new day was dawning.

And then everything fell apart. Just a few months after Shaw announced the ambitious plan, the government backpedaled. Now the experiment is widely regarded as a cautionary tale of what not to do.

The reason why the program crumbled is relatively simple: No one wanted it. New Zealand (and, for that matter, many other wealthy Western countries) took for granted that people would leap at the chance to be recognized as climate refugees. All the government had to do was offer it. But the situation was much more complicated. "The climate migration issue looks like it's much broader than us coming up with a visa," summed up Green Party MP and spokesperson Golriz Ghahraman, herself a former refugee. "Tuvaluans want to continue to be Tuvaluans." [18] In fact, when asked, Pacific Islanders' main ambitions are not to migrate more easily to New Zealand or another country. They want to determine their own future, and more often than not want to be able to stay in their homes. Just a decade previous, the regional Pacific Islands Forum endorsed a document called the Niue Declaration on Climate Change, which among other things recognized "the desire of Pacific peoples to continue to live in their own countries, where possible." [19] This is one reason why even large-scale, coordinated planned relocation efforts meet pushback and are not always successful.

Moreover, even when people do want to leave or acknowledge that doing so is in their best interest, they want to do so on their own terms, and not as refugees. Unfortunately, the term has become stigmatized, denoting a victim who has succumbed to an unfurling crisis around them. "I have never encouraged the status of our people being refugees," Kiribati's President Anote Tong said in 2014, several years before New Zealand unveiled its proposal. "We have to acknowledge the reality that with the rising sea, the land area available for our populations will be considerably reduced and we cannot accommodate all of them, so some of them have to go somewhere, but not as refugees."[20] In the popular imagination, refugees are at the mercy of their charitable hosts, not active members of their new communities. They are rag-clothed families on desperate journeys across the Mediterranean, penniless children crawling through jungles to flee massacres, people on the last train out of their besieged hometowns, hoping only for the graciousness of strangers to take them in. Refugees have become objects of pity. Who would volunteer to be that?

Six years later, Australia improved upon New Zealand's idea. In late 2023, the country's leaders announced an agreement with tiny Tuvalu allowing up to 280 Tuvaluans to migrate to Australia each year under a new type of visa category that allowed them to work, study, and live in the country, but not as refugees. Tuvaluan Prime Minister Kausea Natano called it "a beacon of hope, signifying not just a milestone, but a giant leap forward in our joint mission to ensure regional stability, sustainability, and prosperity."[21] In a joint press conference, Australian Prime Minister Anthony Albanese offered similar praise, calling it "a groundbreaking agreement." Known formally as the Australia-Tuvalu Falepili Union (*falepili* is Tuvaluan for "good neighborliness"), the deal "will be regarded as a significant day in which Australia acknowledged that we are part of the Pacific family." To underscore the sentiment, he wore a bright blue patterned short-sleeve shirt and an elaborate puka shell necklace.[22] While 280 people per year is small in the grand scheme of things, it is nonetheless a significant chunk of Tuvalu's total population of just about 11,200. Theoretically, the plan could pave the way for the entire country to be depopulated in forty years. "This was new in the sense that it was framed as climate change and future impacts that Tuvalu might experience," said Jane McAdam, perhaps the world's leading scholar on the legal intricacies of climate migration. "This is meant to be a voluntary arrangement. Nobody is forcing people to move nor is this akin to a refugee-type protection visa; it's very much about voluntary migration."[23] Natano's administration in Tuvalu had spearheaded the effort.

Again, there was a problem. And again it came not from the destination country (Australia in this case) but from the climate-vulnerable country, Tuvalu. "There's consternation that people weren't consulted about it," McAdam told me. For many islanders, the deal felt foisted upon them. And also, as part of the agreement, Australia got de facto veto power over all of Tuvalu's future international security agreements. This being the Pacific, the elephant in the room was China. Tuvalu is one of the handful of countries that recognizes Taiwan rather than mainland China, much to Beijing's consternation, giving it outsized diplomatic sway for such a small island state. "Some people were very concerned that effectively Tuvalu was now at the behest of the Australian government in terms of what sorts of arrangements it might enter into," McAdam said. Two months after announcing the agreement, Natano was booted out of office as Tuvalu's prime minister.

Still, the agreement persisted and went into effect in 2024. Visa applicants are selected by random ballot, and thousands entered the first drawing, which took place in 2025. In addition to the visas, Australia also formally committed to continue recognizing the statehood of Tuvalu even if the land itself disappears and pledged the equivalent of $24 million for coastal adaptation projects.[24]

Other countries have proposed even more aspirational experiments. In the United States, Senator Ed Markey and Representative Nydia Velázquez have for years introduced bills to allow the annual arrival of at least 50,000 "climate-displaced persons," defined as anyone who is forced to move "for reasons of sudden or progressive change in the environment that adversely affects his or her life or living conditions," who needs durable resettlement and whose own government is incapable of providing it.[25] This new climate-displacement route would exist alongside and in addition to the U.S. refugee resettlement system. Markey is not an unexpected leader on the issue. Despite being a septuagenarian born just after the end of World War II, the suburban Boston native with a thick New England accent has become one of the United States' most dependably progressive leaders on climate matters. In fact, he probably owes the later stages of his five-decade political career to his work fighting climate change. In 2020, he pulled off an unexpected victory over a primary challenge from Joe Kennedy III, a former congressman and the grandson of Robert F. Kennedy, in a dramatic intra–Democratic Party fight that mirrored Joe Biden's nomination contest against Senator Bernie Sanders. Despite being thirty-four years Kennedy's senior, Markey was successful in large part because of his leadership on the Green New Deal, which made him an icon for a younger generation of climate advocates.

When Markey first introduced the bill in 2019, as the first legislation of its kind in the United States, one of its five cosponsors was Senator Kamala Harris; the following year she would be elected as the nation's first female and Black vice president and tasked with, among other things, addressing the factors driving migration to the United States. The White House for a moment seemed at least moderately intrigued. In a 2021 report, it expressed the first inklings of openness to legislating a process for climate migration: "The United States does have a national interest in creating a new legal pathway for individualized humanitarian protection in the United States for individuals who establish that they are fleeing serious, credible threats to their life or physical integrity, including as a result of the direct or indirect impacts of climate change. This new legal pathway should be additive to and in no way infringe upon or detract from existing protection pathways to the United States, including asylum and refugee resettlement."[26] But the Biden administration took no further action on the issue, and Trump's reelection in 2024 guaranteed that nothing will happen for years, if ever. Still, consider the proposal an opening gambit, an effort by some people in what has historically been the world's largest resettler of refugees to redirect the slow-moving legislative machinery.

What if there were another way? What if we simply looked at the spirit of the law, rather than the letter, and acted accordingly?

*   *   *

The UN Human Rights Committee sits in a dark, musty corner of the international legal landscape. It does not hold regular sessions in a grand courtroom with marble columns and ornate floral bas-reliefs. There are no teams of journalists perched to splash its every pronouncement across cable TV and newspaper front pages. It is simply a group of eighteen human rights experts with four-year terms, none of whom would be recognized on the street. Not to be confused with the better-known Human Rights Council, the Human Rights Committee's job is to monitor how countries are adhering to the International Covenant on Civil and Political Rights (ICCPR), a wide-ranging treaty that forms one vertebra of the international human rights system. The proclamations it issues from time to time attract little notice. It's a wonky group with little authority, which is often stonewalled or outright ignored by governments. In 2019, the committee reported that sixteen countries were years overdue on submitting reports about their compliance with the ICCPR; Equatorial Guinea hadn't bothered to submit a report in thirty years.[27]

Yet in 2020, this quiet group delivered a milestone ruling for climate migrants, determining for the first time that someone fleeing climate disaster could be eligible for asylum.[28] It was a groundbreaking decision, based again on a case in New Zealand. Ironically, it was a case where the person didn't even get to stay in the country. The case derived from Ioane Teitiota, a Kiribati native who was denied refugee status in New Zealand. In 2015, he, his wife, and their three New Zealand–born children were deported back to Kiribati, a place where he said sea-level rise was threatening their existence. Increasingly, jobs were disappearing, and Tarawa, Kiribati's capital atoll, was growing more crowded as i-Kiribati migrants moved there from outlying islands. The coast was eroding away and the ocean regularly spilled over the seawall and onto the land, washing out roads and killing crops. He had arrived in New Zealand on a work visa eight years earlier and worked for much of that time on farms and in greenhouses while his wife found a job in a nursing home.[29] His visa ran out and he was arrested in 2011 when a police offer pulled him over for having a burned-out taillight. He appealed to the courts to extend his stay, first to New Zealand's Immigration and Protection Tribunal and then up through the appeals system. As it would do later, in the 2017 case concerning Tuvaluans, New Zealand's legal system shrugged its shoulders and said Teitiota's situation did not merit him international protection.

Back in Kiribati, he drank collected rainwater, he told the BBC, since the pump on his family compound produced a filthy mess often contaminated by human and animal waste. Holes regularly appeared in the chest-high seawall by his family home. "I'm the same as people who are fleeing war," he said. "Those are who are afraid of dying, it's the same as me. The sea level is coming up and I will die, like them. It will affect my life when the sea takes over my land. It will kill me and my family."[30] He continued to keep up the fight, bringing the case to the Human Rights Committee. Ironically, the committee did not rule that Teitiota himself had had his rights violated by being removed to Kiribati. New Zealand has properly assessed his individual case, weighed the threats against him, and was within its right to conclude that he could safely be removed to Kiribati, the committee said. But the committee nonetheless set a standard for judging how climate change might be considered in refugee cases going forward.

The ruling gets at a tricky legal principle at the heart of

refugee law: non-refoulement. The Refugee Convention's definition of non-refoulement is somewhat narrow, demanding that signatories not "expel or return ('refouler') a refugee in any manner whatsoever to the frontiers of territories where his life or freedom would be threatened on account of his race, religion, nationality, membership of a particular social group, or political opinion," although it includes exceptions for refugees "whom there are reasonable grounds for regarding as a danger to the security of the country in which he is, or who, having been convicted by a final judgment of a particularly serious crime, constitutes a danger to the community of that country." [31] Other legal documents have gradually expanded on the concept, eliminating exceptions and creating an umbrella of protection to evolve with the times. [32] The 1984 Convention Against Torture, for instance, says that signatories cannot "expel, return ('refouler'), or extradite a person to another State where there are substantial grounds for believing that he would be in danger of being subjected to torture." [33] The European Convention on Human Rights forbids someone from being "subjected to torture or to inhuman or degrading treatment or punishment," which does not explicitly mention non-refoulement but has nonetheless been interpreted to include it. [34] In one notable ruling, the European Court of Human Rights determined that that measure prevented the United Kingdom from extraditing an alleged murderer to the United States, where he might be subjected to the death penalty. [35]

The ICCPR also is based on the principle of non-refoulement. Under Article 6, "Every human being has the inherent right to life" and "no one shall be arbitrarily deprived of his life." Under Article 7, "No one shall be subjected to torture or to cruel, inhuman, or degrading treatment or punishment." [36] In its ruling on Teitiota's case, the Human Rights Committee said it was "of the view that without robust national and international efforts, the effects of climate change in receiving states may expose individuals to a violation of their rights under Articles 6 or 7 of the Covenant, thereby triggering the non-refoulement obligations of sending states. Furthermore, given that the risk of an entire country becoming submerged under water is such an extreme risk, the conditions of life in such a country may become incompatible with the right to life with dignity before the risk is realized." [37] In other words, returning someone to a place where there is a real risk that the impacts of climate change could threaten their right to life or right not to experience inhuman or degrading treatment would violate the principle of non-refoulement. Therefore, countries had an obligation to protect them. There it was: a big red line. It was awkward and indirect, but the world nonetheless had something resembling the definition of a climate change refugee and the legal framework for protecting them.

The legal discussion is only in its infancy. Even if it were to be adopted widely, this line of legal thinking would only prevent countries from deporting people back to climate-affected areas, not proactively allow people currently living there to escape. And it sets a high bar; as Teitiota's own case shows, governments are by no means bound to accept anyone who wants to stay in the country. Precedent-setting or not, Teitiota got precisely nothing for his troubles. "Forgive my ignorance, but to be frank, I'm quite disappointed with the outcome of my case," he said after the ruling was issued. "It's still the same as before—I'm still worried about my family [because of] climate change . . . the sea-level rise, the drinking water is not good . . . [and] I'm still yet to find a job until now." [38] If a man on one of the world's most climate-affected islands whose house was quite literally sinking into the sea didn't merit protection, who would? Moreover, the Human Rights Committee's rulings are nonbinding. The panel does not set law for any national government, and legal volumes are filled with countries wantonly ignoring advisory rulings against their own policies. No matter how far-reaching its rulings may be, the committee has zero way to enforce its decisions.

Still, it's a start. A vocabulary for lawyers to work with. A toehold for future legal climbs. That might not be much, but it is a necessary something. And slowly, the image of a future legal system is beginning to come into view. In late 2020, several months after the Human Rights Committee's ruling on Teitiota, a court in France overturned a deportation order for a Bangladeshi man with a respiratory disease, deciding that he would have faced serious repercussions and possibly even death due to pollution in his native country.[39] The forty-year-old man, whose name was not made public, had lived in France since 2011 and held a temporary residence permit based on his need of medical treatment, but doctors later determined that he could be adequately treated in Bangladesh, leading to an expulsion order. In its ruling, a court in Bordeaux rejected that claim, asserting that the environmental conditions had to be taken into account. The man's lawyer, Ludovic Rivière, called the ruling a "shortcut" to climate refugee status which recognized that without accounting for climate issues in migrants' origin countries, lawyers would be "reasoning in a vacuum."[40]

Parts of these arguments resemble legal thinking that has been around for years. Since 1990, the United States has offered Temporary Protected Status (TPS) to hundreds of thousands of immigrants whose origin countries were deemed to be undergoing crises that made it unsafe to return. TPS was first used for immigrants from El Salvador, which was caught up in civil war. Since then, it has been used to grant liminal legal status to immigrants from more than two dozen countries. But like other non-refoulement policies, TPS is designed only to prevent immigrants from being deported to dangerous places, not protect those in the middle of a crisis who are looking to escape. The rub is that recipients have to already be in the United States at the time the designation is declared. And as the name implies, it is temporary, designed to grant legal status only for a few years at a time, threatening them with the prospect of being eventually deported to their crisis-stricken origin country. Although some recipients, including many Salvadorans, have had their status renewed over and over again, the Trump administration has raced to withdraw the protection for hundreds of thousands of Afghans, Haitians, Venezuelans, and others, underscoring the system's precariousness.

If these examples tell us anything, it's that the legal solution to climate displacement is a lot more complicated than we'd like to believe. For one, it is often difficult to tell precisely when someone is moving in response to climate change versus when they are moving for other reasons. Climate change works in mysterious ways. But perhaps for that very reason, it is necessary to think of climate migration not as a wholly new phenomenon but as one iteration of the complex process of human migration.

5

Empty Bellies and Empty Pockets

The Dry Corridor (*Corredor Seco*) of Central America runs like a bulbous head of a bird with its neck in the northwest of Guatemala and its beak folded around Nicaragua's Lake Cocibolca, passing through El Salvador and Honduras on the way. It stretches around 1,000 miles long and is up to 250 miles wide at its fattest, covering more than 40 percent of the combined land area of the four countries. More than 10 million people live in the Dry Corridor, many of them in small rural communities with little plots of land where they grow food either for themselves or to sell.[1] These farmers barely eke out a living even in good times, but there have not been too many of those recently.

It is a belt within a belt, a semiarid strip of heat straddling the length of Central America, which is blocked on either side by mountain ranges that prevent moisture from the ocean, the climate scientist Paris Rivera told me in his office in Guatemala City's Mariano Gálvez University. On one computer monitor, we watched live as a hurricane gathered force in the Pacific. On another, he walked me through Guatemala's rapid environmental change, with temperatures heating up and patterns of precipitation growing more extreme. Over the last twenty years, the average temperature has risen very quickly, he said. From virtually no change between 1970 and 2000, average temperatures climbed nearly 1°C (1.8°F) between 2000 and

2022. More importantly, overall rainfall has increased slightly, but it's been coming in more extreme episodes; long periods of dry are followed by intense storms. When it rains, it truly pours.

As the bridge between continents, Central America seems to get battered by all the world's climatic ills. Hurricanes come from both the Atlantic and the Pacific, and, during so-called crossover storms, start in one ocean and end in the other, battering the full width of the region. Meanwhile, drought wracks the interior Dry Corridor. These two phenomena can intersect in tragic ways, making the ground too dry and brittle to absorb water, so that when it rains it causes flooding and mudslides that impact communities who live on mountainsides. These threats have been exacerbated by deforestation and farming techniques that erase the topsoil with each new growing season. The fungus that causes coffee leaf rust, a pernicious disease known locally as *la roya* and that withers coffee plants, has spread across Latin America on the backs of rising temperatures, wiping out vast plantations. Cycles of El Niño and La Niña, which in Central America correspond to periods of less and more rain, respectively, have grown more intense. Things have always varied from year to year, but now the pendulum is swinging wildly. "What happened is climate change coming on top of the natural variability to make the extremes more extreme," another climatologist, Edwin Castellanos, told me.

I came to Guatemala because there is perhaps no place where the nuanced and complex intertwining of climate change with economic, political, and historical factors more visibly shapes international migration. Here, climate change has combined with long-running poverty and development challenges as well as social and political marginalization. It has also dovetailed with a history of migration that predated the climate change era but which made international movement seem especially feasible. Formally classified as an upper-middle-income country, Guatemala has average incomes about twice those of Bangladesh, meaning more people have the resources to move, and the relative closeness of Mexico next door and the United States a little farther on has made moving less logistically difficult than it might be in small Pacific Islands. Climate change alone is not forcing people to move, but it does not occur in a vacuum. When it combines with other forces in particular ways, emigration might seem like the best option.

Driving out of Guatemala City and toward the Dry Corridor, some of the region's extremes became obvious. The capital city itself is modern and big—the largest in Central America— with wide avenues, freshly paved highways, and buildings of glass and steel. But upon leaving the valley in which the city lies, the road winds around vertiginous cliffs of the country's volcanic arc, where lush greenery gives way to sharp drops into ravines below. The dizzying shifts in altitude create a series of microclimates in which just a short trip up- or downhill can yield a 10- or 20-degree change in temperature.

As we got close to the Honduras border, our car crossed a newly constructed bridge stretching over an angled gorge and then turned left, off the paved road and onto a steep path stretching precipitously skyward. It was perhaps the steepest road I've ever seen, the packed dirt crosscut with rivulets and canyons washed out by stormwater. The only other cars here seemed to be other Toyota Hilux pickup trucks, most of them with beds carrying entire families and cases of cellophane-wrapped soda bottles, rice, and other groceries. Every now and then we passed a person hiking up or trekking down, sometimes carrying bags on their shoulders. Walking all the way to the bottom of a valley could take half a day, given the slow pace of moving up and down the mountain. We passed fields of bananas, beans, corn, and coffee—the first three primarily for eating, the last for selling—and finally came to a stop in the tiny village of Barbasco, at the home of Consuela, a forty-year-old Ch'orti' Maya woman with kind eyes. Skinny dogs, their ribs and hip bones peeking through weathered fur, milled about and curled up in the dirt. A couple of small children played nearby, one running a car up and down the miniature gullies of cracked earth. Lately, Consuela and the rest of her family had been sleeping in a detached kitchen, a small tin-roof building with a solar panel on top where her sixty-seven-year-old mother was busy washing clothes and preparing dinner. That's because the actual house, a two-room clay-sided structure next door where she had been living for the last fifteen years, was cracking in two. Storms and hurricanes had split the earth, creating a six-inch gash in the dirt floor where one end of the building was beginning a slow march toward the edge of a cliff. A crack had worked its way into the wall, creeping up the side of the house like a sinister vine. The house has been slowly drifting toward the edge of the mountain, Consuela said, inch by inch and bit by bit, as the ground underneath started to give way and slide downhill. It's a common sight on these hills, where coffee and corn plants perch precariously on steep slopes that threaten to give way with the next storm.

Since the late 1990s, there have been a lot of big storms. Hurricane Mitch, in 1998, swept across Guatemala from east to west before turning up into Mexico, leaving devastation that was even more pronounced in Honduras and Nicaragua. Tropical Storm Agatha killed nearly two hundred people in Guatemala in 2010.[2] Most recently, in late 2020, Hurricanes Eta and Iota ravaged the Caribbean coast, with significant damage stretching more than one hundred miles inland that prevented some people from returning to their homes for multiple years and making it harder for residents to prepare for future storms.[3] In the small town of San José La Arada, for instance, the hurricanes tore down the banks of a nearby stream; two years later, when Tropical Storm Celia rolled in, this erosion allowed the water to burst the banks and rush through the town for the first time in memory.

Similar dynamics happen in more populous areas. Rain causes erosion that can lead to massive sinkholes and wipe out bridges and other infrastructure. Guatemala City is built in a valley, crosscut by steep ravines (*barrancos*) that separate neighborhoods and act as natural urban barriers. Technically, the walls of these ravines are not zoned for people to live in, but because they are essentially no-man's-land they can seem inviting to newcomers with little money in their pockets who are looking for a place to rest their heads. As a result, many of the city's traffic-clogged streets run under concrete blocks jutting from the green of the gorge like a toothy grin.

Informal settlements like these are always vulnerable to the impacts of climate change, given their haphazard construction and crowded conditions. But when built on a steep gradient in a country that faces routine hurricanes and extreme rainfall, they can be ground zero for deadly landslides. In 2015, more than 270 people were killed in a landslide on the southeastern outskirts of Guatemala City, in a place called Santa Catarina Pinula.[4] The mudslide happened in the evening, when houses were full of families. Loosened by rain, rocks and earth tumbled down the side of a ravine, crushing dozens of structures in the town of El Cambray II. Some houses were buried under fifty feet of earth. Victims lived long enough to be able to text pleas of help to their loved ones before succumbing to their injuries or asphyxiating.[5] A similar dynamic has played out in Brazil, where informal hillside favelas are under constant threat of falling victim to mudslides.[6] And indeed, the threat is facing any place where cheap structures rest precariously on hillsides that face persistent seasonal deluges.

At the very least, storms, mudslides, and hurricanes have the ability to capture global attention. There is a chance that, after disaster strikes, governments and organizations will invest in protective structures to prevent tragedies from happening again. But no less serious for Consuela are the changing seasons, which have upended a previously meticulous pattern of planting and harvest that ought to allow her to plant corn and beans one after the other on her small plot of land. Because the seasons have been changing, the careful calendar has been thrown into disarray. When I visited in late August, it should have been time to start harvesting the corn and preparing to plant beans, but the harvest seemed still weeks away. As a result, she often goes hungry, as do many of her neighbors. Across Chiquimula, the province in which Consuela lives, more than 50 percent of the population faces chronic malnutrition.[7] In bad times, there can be little more to eat than a few salted tortillas or cooked corn husks.

On the other side of the mountain, a fifteen-minute drive from the Honduras border, I met José, a soft-spoken thirty-year-old with close-cropped hair and a round face. It was markedly hotter at José's home than at Consuela's. He too often had problems feeding his family, including four children, one just five months old. The impact of drought for him was less immediately dramatic than the torrential rain that had ripped at Consuela's home. Its effects may take multiple years to become clear, making it "the silent catastrophe," said Castellanos, the climate scientist. Drought can cause wildfires, the brittle undergrowth turning to kindling in the heat. But in Guatemala, where about 30 percent of the country works in agriculture, the biggest impact of drought is crop failure, which leads to poverty and hunger.[8]

The Dry Corridor is not the origin of the largest number of migrants from Guatemala. Because it has historically been rural and not conducive to large-scale agriculture, it's relatively sparsely populated. The province of Chiquimula accounts for only about 3 percent of the total national population of 14.9 million, according to the 2018 census.[9] But it offers perhaps the most dramatic example of how climate change can aggravate economic conditions, pushing people to the brink of survival. And as climate change accelerates, it could just be Guatemala's future. In coming decades, "you will not need to travel to Chiquimula to see the dry forest; you will see it right here in the city," Castellanos told me. The same may be true of the vast northern area of Petén, a densely forested area bordering Mexico and Belize that is home to the ancient Mayan ruins at Tikal. Semiarid lands such as the Dry Corridor may come to resemble deserts, and current forests may turn into scrubland.[10] The Dry Corridor may soon become a desert highway.

Facing conditions like these, it should not be surprising that many people are looking to leave. When asked, Central

American migrants have recently tended to say they left for
economic reasons. But increasingly, it has become impossible
to disentangle economic from environmental and other factors, especially for rural farmers whose livelihood comes from
the earth. Consuela's seventeen-year-old son, Francisco (not
his actual name), left home a year before I met her, taking
the dangerous and expensive trek to the United States, where
he entered without authorization and settled in New York. It
was a difficult decision to make for him to leave, and one that
Consuela stills seems conflicted about. She expressed regret
at the move, worrying that the cost of migration was so high
and had yet to pay sizable dividends. More than two-thirds of
the $300 Francisco sends back each month is diverted to pay
off the debts to the coyote, who accepted Consuela's land deed
as collateral for the nearly $6,000 trip (plus significant inter-
est). Clearly, Francisco's move was a gamble on overcoming
the steep challenges that the family faces in Barbasco, some
of which have been caused by extreme storms and poor agricultural conditions. But Consuela herself is in the midst of her
own sort of migration, of a much smaller nature. To escape
the house slipping off the cliff, she has begun building a new
structure fifty yards uphill, in a flat area set back from the lip
of the mountain. It's a pricey move; materials alone have cost
her more than $1,000, which is a cost she can ill afford.

On its face, Consuela's problem is not obviously related to
the climate. Unlike Pacific Islanders, for whom the rising seas
are a clear and existential threat, or Bangladeshis in the Sundarbans whose houses keep falling into the river, the changing environment is more abstract for her. When we spoke,
she didn't talk in the language of climate shifts and average
annual precipitation levels. The Americas have gone through

patterns of La Niña and El Niño for years, and there have been droughts and storms of one kind or another for as long as the earth has existed. At its root, Consuela's problem is that she doesn't have enough money. But scratch just beneath the surface, and it's clear that climate change is making matters worse. If Consuela were not poor, her problems might not be so bad. But one reason she is poor is because she depends on a very particular kind of climate to thrive, and that climate has been changing. She's hungry because she's poor; she's poor—or at least poorer than she might be otherwise—in part because of climate change. And throughout history, one way to get out of poverty has been to leave home.

Groups such as the World Food Program have found a clear connection between hunger, climate change, and emigration. In fact, nearly half of all Central American migrants interviewed by the World Food Program in 2016 lacked the necessary amount of food.[11] Central Americans experiencing declining economic conditions or a worsening standard of living are likely to have recently experienced a natural disaster.[12] Drought can have a direct impact on migration. Between 2012 and 2018, Central Americans in regions with particularly dry weather were 70 percent more likely to emigrate to the United States than those in places with more typical weather.[13] Researchers in El Salvador found that extreme temperature was strongly correlated with heightened emigration from places where people mostly depend on growing corn.[14]

Many people will try to account for changes in their local environment by altering their habits, such as digging into their savings or cutting back expenses. But that may not be enough, and they may eventually have to leave home as a last resort. For others, a worsening economy may be proof that now is the time to go, so they will speed up their plans to get out before things get even bleaker. Omar Ramirez Rivera, a development worker in Central America, told me he sees a direct correlation: "The less food, the more migration." Although only to a point, he maintained. When things are going well, men are more likely to migrate, often to look for work in the United States. When things get bad, women will use the last of their savings to follow. But when things get really bad, like during a period of severe drought or after a major storm, everyone stays put, because nobody has the money to leave.

In broad strokes, the situation in Guatemala is similar to that in dozens of other places. Poverty in the rural countryside and urban slums combines with long-standing political marginalization, lack of opportunity, and the existence of an established route to a brighter future either out of the country or to somewhere else internally. Put it together and the odds increase that someone will leave. Change the names and the specifics, and this could be Nigeria or India or Russia or the United States a century ago. All around the world, people want to make a better future for themselves and for their children. For all of human history, one way to do that is to go someplace else.

So at what point are the people in the Dry Corridor climate migrants, versus migrants generally who want a better future, like my own ancestors and, almost surely, some of yours?

It's not so clear. We like to think in discrete concepts. Apples are not like oranges. Cats are not like dogs. This person is a refugee and this one is looking for a job and this one wants to be closer to her grandkids. But the truth is much more complicated. All of us are driven by a million different motivations, influenced by an infinite number of emotions and desires and fears. We put labels on our lives and fit them into established narratives and categories to simplify something that is inherently messy. We pass laws and implement policies to bring order to chaos. But the world does not always fit into these neat boxes, and something or someone usually gets left out.

Another way to approach the question about whether people leaving drought-affected places in Guatemala qualify as climate migrants is to ask whether these same people would be moving in a counterfactual universe in which human-caused climate change did not exist. Some of them probably would, because there is a history of migration from Guatemala to the United States that stretches back to the late 1970s, spurred by domestic unrest, political violence, and marginalization. Small numbers of Guatemalans have migrated to the United States since at least the 1800s. More recently, President Jacobo Árbenz's ouster in a U.S.-backed coup in 1954 led to a brutal counterinsurgency campaign and thirty-six-year civil war during which the government waged a terror campaign against leftist insurgents, many of them Maya, who account for slightly more than 40 percent of the national population.[15] Many persecuted people and others fled the country. During the coup and the army's actions in the 1960s, most emigrants went right next door to Mexico, but emigration ramped up and more people started heading to the United States in the 1970s and the 1980s, as the conflict became more intense.[16] Between 1981 and 1983, government forces destroyed more than 440 villages, according to their own tally, mostly Maya in the Western Highlands bordering southernmost Mexico, leaving 150,000 dead or disappeared and prompting 200,000 people to flee to Mexico and 1 million to be displaced elsewhere within Guatemala. The government's target during this time was rebel groups aligned with the Unidad Revolucionaria Nacional Guatemalteca (Guatemalan National Revolutionary Unity), but many peaceful civilians were caught in the cross fire, presumed to be silently supporting the rebels or simply just in the wrong place at the wrong time. In all, more than 200,000 people were killed or disappeared, according to the UN-backed Commission for Historical Clarification, which in 1999 determined that government forces had carried out a genocide against the Maya.[17] "Faced with several options to combat the insurgency, the State chose the one that caused the greatest loss of human life among non-combatant civilians. Rejecting other options, such as a political effort to reach agreements with disaffected non-combatant civilians, moving of people away from the conflict areas, or the arrest of insurgents, the State opted for the annihilation of those they identified as their enemy," the report declared. Eighty-three percent of the victims who could be identified were Maya.

Land disputes were a major theme of Guatemala's civil war, which was among other things prompted by marginalized and Indigenous people disenfranchised after decades of being sidelined by corporate powers and the elite with ties to international interests. Officials linked to the United Fruit Company—a sprawling corporation that inspired the term "banana republic"—were major backers of the U.S.-supported coup against Árbenz, who had, in their eyes, committed the crime of pushing to redistribute ownership of land in his country. The company was Guatemala's largest landowner and has been described as operating a state within a state, owning the country's telephone and telegraph facilities, running the major Atlantic coast harbor, and through a subsidiary owning virtually every mile of Guatemala's railroad track.[18] Árbenz's reform sought to divide up more than half of its holdings, offering payments based on what the company had claimed in tax returns its land was worth, which corporate officials suddenly decided was a massive undervaluation.[19] In earlier generations, land-grabbing was a theme when the coffee industry swept into Guatemala and El Salvador in the 1800s and, in Guatemala, turned Indigenous communities into rural proletariat and peasants.[20] Like in Bangladesh, it is only a slight oversimplification to say that many of the structural tensions in Guatemala can be boiled down to who owns what land in which places. Even though the war is now over, disputes about land are still an undercurrent of the country's internal challenges. Many rural farmers face ongoing abuse from corporate powers that seek to buy and claim ownership of fertile and resilient lands. As a result, communities like Consuela's have ended up living in remote places that are particularly vulnerable to the impacts of climate change, which can have devastating effects on their ability to feed their families.

Emigration during the war and genocide was part of a broader trend of migration from Central America to the United States in the last decades of the twentieth century, alongside migration from neighboring El Salvador and Honduras. Across Latin America, the 1980s have been described as a Lost Decade (*La Década Perdida*), in which national debts and inflation combined to plunge multiple countries into economic tailspins. Pushed by international groups like the International Monetary Fund, countries slammed the brakes on government spending, adopted liberalized open markets, and focused on importing and exporting goods rather than producing products for themselves. Meanwhile, advances in medicine, infrastructure, and living conditions helped rural populations grow tremendously, but new children were born into areas without enough jobs or educational opportunities for them to thrive. Over the 1970s, the number of small farmers in Guatemala increased by 54 percent, but the land area for them to use grew by only 6 percent.[21] In some parts of the Western Highlands, the population doubled between 1950 and 1980.[22] Something had to give. As economies plummeted, moving abroad became increasingly attractive.

Meanwhile the United States was undergoing a dramatic economic transformation of its own, as Reagan-era changes empowered corporations over labor unions, supported offshoring of manufacturing jobs, and increased investment in financialization over production. The United States at this time was transitioning into an economy built around providing services rather than producing goods. For workers, this tended to mean an economy split in two: well-educated professionals in white-collar jobs such as lawyers and financial advisers, and positions with lower wages, less security, and fewer benefits for those with less education. Hiring unauthorized immigrants was also not illegal before 1986. So as working conditions and pay got worse, companies increasingly turned to unauthorized immigrants to fill the holes in the labor market—and drive down wages further. In 1986, passage of the Immigration Reform and Control Act allowed about 2.7 million unauthorized migrants from Central America and elsewhere to gain legal status, the United States' last major so-called amnesty program. Nearly 50,000 Guatemalans received green cards due to the law (about 2 percent of all beneficiaries), granting a toehold not just to them but also to family members who might be able to follow them through legal pathways.[23] U.S.

immigration law allows family members of green-card holders and citizens, which people with green cards can become, to apply to immigrate in a process that is sometimes derisively referred to as chain migration. In 1990, the United States offered the liminal protection known as Temporary Protected Status (TPS) to Salvadorans, and in subsequent years would do the same for Hondurans and Nicaraguans, but not Guatemalans. In any case, legal status offered immigrants a chance at a better job in the United States, which meant a better lifestyle and more money that could be sent back to their countries of origin. And it allowed immigrants to travel back and forth from the United States to Central America to see friends and family. Returnees often brought back money, TVs, and other things that made their lives in the United States seem attractive, encouraging others to leave Guatemala. In the process, the back-and-forth movement fostered the growth of a dedicated industry of buses, hotels, and other migration-related services in Mexico, all of which made it easier for new travelers to make the journey.

Enforcement by authorities in Mexico and at the U.S. border ramped up in the late 1980s and 1990s, most clearly when the United States passed the Illegal Immigration Reform and Immigrant Responsibility Act in 1996, which increased the penalties for unauthorized immigration and led large numbers of immigrants to be deported. Mexico's increasing focus since 2001 on policing its southern border has had a similar effect. But by then the die had been cast. There were just 5,400 Guatemalan immigrants in the United States in 1960, but the number shot up to 63,100 in 1980, 480,700 by 2000, and 1.3 million by 2023.[24] Add to that the more than 1 million children and grandchildren of Guatemalan immigrants who were born in the United States, and it's clear to see how Guatemalans became one of the fastest-growing immigrant communities in the country.[25]

Yet as the increasingly restrictive policies suggest, the situation in the United States was not welcoming toward Central Americans, especially those fleeing conflict. Historically, Central Americans and Mexicans have been disproportionately unlikely to receive asylum in the United States. From 1983 to 1990, less than 2 percent of Guatemalan asylum seekers and less than 3 percent of Salvadoran asylum seekers received U.S. protection.[26] The official line was often that they were not facing true persecution in their origin countries but were simply heading to the United States to find work. The evidence provided for this claim was that Central American migrants were not seeking protection in Mexico. However, the notion that migrants were fleeing persecution was also at odds with Washington's support for the right-wing governments in Guatemala City and San Salvador, as part of the Reagan-era effort to quell leftist movements in Central America. The Reagan administration saw these governments as allies in the broader Cold War battle against communism and endorsed their bloody crackdown as one front in that fight. Migrants fleeing Guatemala and El Salvador simply couldn't be fleeing persecution, Washington's narrative went, because the governments were the good guys. Acknowledging the threats they faced threatened to undermine the entire logic of U.S. policies in Latin America.

Washington's narrative was ultimately proven wrong. In settling a class action lawsuit in 1991, *American Baptist Churches v. Thornburgh* (often referred to as simply the *ABC* settlement), the Justice Department tacitly acknowledged that officials had failed to properly consider asylum applications from Guatemalans and Salvadorans, thereby withholding protection from many people who probably should have qualified. The reason was clear: Granting asylum to a large number of people from these two countries would amount to a recognition that they were in fact run by brutal autocrats, not freedom-loving heroes. In settling the case, the government allowed about 300,000 asylum seekers from these countries to reapply for protection and remain in the United States.[27]

All this history helped build what researchers describe as a culture of migration, in which leaving your origin country—sometimes temporarily, sometimes seasonally, and sometimes for good—has become commonplace. Coupled with the rise in disasters and other impacts of climate change, the prospect of moving out of Guatemala can be appealing. It's similar to how people in other contexts might think about going to college or getting married and having kids. Not everyone does it, but basically everyone knows someone who does. Driving through rural Guatemala, my driver and local guide, Conrado, pointed down the hill to a number of concrete-slab houses that seemed noticeably empty. They were *pueblos fantasmas*, he said—ghost towns where there are more houses than people. The residents had left, abandoning their property. The vast majority of those leaving the country go to the United States, although some settle in Mexico, Belize, or elsewhere.

Across Guatemala, El Salvador, and Honduras, 43 percent of households said in a 2021 survey that they would like to migrate permanently within the next year. That by no means suggests that all of them did in fact move; there is a big difference between saying you'd like to do something and actually doing it. But those social expectations and networks make it easier for people to leave in the face of droughts or storms. And when a disaster comes or a crisis like hunger strikes, that can be a powerful motivator for people to transition from thinking of migrating to actually embarking on the journey. Among people experiencing food insecurity, like many of those in Chiquimula, 23 percent of survey respondents said they were making concrete plans to migrate, compared to just 7 percent of those who were food secure.[28] There are of course many other factors that affect whether or not someone will leave their home and take a dangerous journey of more than a thousand miles to a place where they likely don't speak the language, navigating organized crime and armed border guards in two foreign countries along the way. Having a friend or family member abroad is one major factor in whether people leave.[29] They also have to be able to afford a trip, which would be far out of reach for many poorer families.

Thinking back to how migration works generally, we have here push factors in Guatemala (conflict, poverty, climate change) and pull factors in the United States (the promise of more money and economic liberalization that made Guatemalans attractive to U.S. corporations). Individuals considering migration must weigh their means (money they have or can borrow to finance the journey, friends and family members in their destinations who can support them) with their aspirations (the prospect for a better life somewhere else, weighed against what they would lose if they left).

The last few decades have seen ebbs and flows in migration, yet the increasingly restrictive policies in the United States and Mexico have meant that migration was often irregular or illegal. Aside from Mexicans, Central Americans have historically been the most commonly apprehended group of migrants trying to cross the U.S.-Mexico border without authorization, and about one in every ten unauthorized immigrants in the United States is estimated to be from Guatemala.[30] Meanwhile, the odds of receiving protection in the United States remain slim, even years after the Cold War and the 1991 *ABC* settlement. Between 2001 and 2021, just 19 percent of Guatemalan asylum seekers received protection, as did 15 percent of Mexicans, 18 percent of Hondurans, and 20 percent of Salvadorans. In that time more than two-thirds of Chinese applicants and 73 percent of Nepalese received asylum.[31] As always, there are many other factors at work, including the fact that the geographic proximity of Mexico and Central America to the United States reduces the cost for asylum seekers, meaning that even if they have a minimal chance at sanctuary, more people might try their luck. But whatever the cause, it is clear that people from this region—which, in addition to climate change, has undergone a tremendous amount of conflict and economic strife in the last half century—have a particularly hard time receiving protection. Partly as a result, more than half of all Central Americans in the United States are believed to lack legal status, including those covered by a smattering of liminal legal protections such as Deferred Action for Childhood Arrivals (DACA) and TPS, which temporarily protect people from being deported and allow them to work but don't offer a way to secure a green card or become a citizen.[32] They have also borne the brunt of some of the United States' most anti-immigrant rhetoric. Regardless of their background, Donald Trump and other nationalists have repeatedly referred to Central Americans as gang leaders, drug dealers, and rapists.

The growing antagonism by the U.S. government has meant

that travel can be dangerous and expensive. In normal times, illicit travel from Central America to the United States—including minibus drivers, flophouses, guides, and bribes to police—can easily run north of $10,000.[33] But people running away from environmental devastation face a paradox, in that catastrophic disasters and deteriorating environments have a tendency to wipe away the livestock, crops, and other assets that make up their wealth, meaning that the very poverty they may be trying to escape can also prevent them from leaving. One woman in Chiquimula told me her husband paid more than $20,000 to hire a coyote to get to the United States, more than several years' salary for an average Guatemalan and typically far out of reach for poor farmers from the Dry Corridor.[34] Virtually no one has that much money on hand, so individuals will often rely on loans from family or friends, go into debt, or put up their land to cover the cost. It's a gamble, and it doesn't always work out. Coyotes don't offer refunds to the thousands of migrants who get caught and deported. They also charge interest, so that $20,000 is only the beginning of how much the family will pay.

So are people leaving the Dry Corridor climate migrants or not? Almost surely, some of them would be migrating from Guatemala to the United States even if the seasons were predictable and the weather was always mild. The same is true for the large number of rural-to-urban migrants, whose journeys tend to be cheaper and easier than international travel, especially as the United States and Mexico turn more punitive. Slightly more than half of Guatemalans live in a city, more or less similar to the share worldwide, and the number has dramatically increased each year.[35] We cannot say that all or even most of this migration away from the Dry Corridor is purely a function of climate change.

But climate change has very clearly compounded deep-rooted structural forces and aggravated the factors that make migration more attractive. Within Guatemala, environmental devastation in rural areas will continue to drive urbanization. By 2050, there could be more than 2 million climate migrants moving domestically across Central America and Mexico, according to the World Bank; cities will be a major target.[36] Guatemala City, in particular, already the largest city in Central America, is forecast to be a major destination for many migrants leaving the Dry Corridor and the Pacific coast.[37] Guatemala's history has created a situation in which climate change may not be the sole or even primary reason why people are migrating, but it is an inextricable and increasingly pertinent part of the story. With all the other structural forces in place, sometimes people just need one more push to get out the door. In the Dry Corridor, the dying fields can be the final straw.

6

Migration Can Be a Solution— for People Who Can Get Out

In Guatemala, outside the town of Jocotán, in a house hidden from the main road by a thin wall of vegetation, I met Elena, a slight thirty-eight-year-old with bright eyes and dark hair that was just starting to show the first hints of gray. Elena has seven kids whom she spent most of her time caring for, while her husband found unsteady work as a for-hire farmer. Her husband's job paid enough to get by, just barely, but the family struggled to travel to see a doctor for their five-year-old daughter, who had an undiagnosed heart issue. The eldest daughter, who was nineteen, had been going to school but dropped out during the COVID-19 pandemic because they could no longer afford the $40 per month for books, her uniform, and other costs. Meanwhile everything was getting more expensive, Elena complained, and it often did not rain enough to yield a fruitful harvest. We met in a neighbor's dirt-floor compound, where chickens and ducks pecked at a trash heap and muddy patches of ground. Behind me, tortillas smoked on the stove in the detached cinder-block kitchen. Nearby, cars and trucks rumbled down the main road heading to the Honduras border a half hour away.

When I asked Elena about the prospect of going to the United States, a shy smile crept across her face. Her husband talks about it, she said, but she knows it's just a dream. It would cost thousands of dollars to hire a coyote and make the trip, and the only way they could raise that kind of cash would be to put up their land as collateral. Some of her neighbors have made that bet, and sometimes it has worked out, but not always. Migration would be a huge risk. It could take weeks or months for Elena's husband to travel and establish himself in the United States, assuming he could even get in, during which time Elena would have no source of income. If things didn't work out and her husband was deported or couldn't pay off the coyote's fee plus interest—or worse, if he was injured or killed en route—their land could be swiped from under them, leaving them even worse off than they already were. She would go in a heartbeat, she said, if only it were realistic. But it wasn't.

We might say that Elena is trapped. Emigrating to the United States would almost surely be transformational for her and her family, even if just one member could establish a foothold there. She would no longer need to worry about going hungry. It could put her family on a path of upward mobility, with better access to health care and education. Elena's children would probably have a markedly more comfortable life than her own, which is the desire of all parents worldwide. This was exactly the story of hundreds of millions of Americans whose ancestors scraped together their savings to come to the United States, where they suffered abuse and worked hard, benefited from and contributed to its economic growth, and in a few generations had descendants with a profoundly different quality of life. Instead, she and her family are stuck in a rural farming community where the land is drying up and falling apart and where the prices at the market are climbing ever higher.

Migration would also probably be good for Nur, the woman I met in Bangladesh who had been displaced by floods nine times and spent two years living on her boat. She is bright and entrepreneurial and would surely thrive if she could move to a city or someplace where she could earn money doing something other than drying tiny fish in the sun day after day. Because she was poor, she was stuck in a place where every cyclone threatened to wipe away everything she owned, just as it had done before.

In the long run, "trapped populations" may be the worst victims of climate change. Migration costs money and can be complicated and, if traveling internationally, usually illegal. Leaving might enable people like Elena and Nur to find better-paying jobs elsewhere and send back money that could help protect their homes and families against encroaching climate change.

Yet for a million reasons people stay in place, even if doing so is dangerous. Many of them cannot leave. When disaster strikes, people with disabilities, the elderly, and the poor tend to be less likely to be able to evacuate, and therefore account for an outsize number of fatalities. When Hurricane Katrina hit the United States, for instance, about half the dead were seventy-five years old and older.[1] Moreover, there is probably no way that Elena could get to the United States legally, and laws can be difficult things to break, especially when they are backed by the full weight and force of the U.S. government. It also often takes connections to be able to move, which many people do not have. If Elena had a cousin or friend in the United States who would help her out—tell her whom to call, where to stay, how to get a job—her family might have an easier time. Without this social capital, she's facing an uphill climb.

And of course, it is about one thousand miles to get from Guatemala to the United States, and even farther for people from Asia or the Pacific Islands. Deserts and oceans are physically difficult to cross, and often deadly. It is a sad fact that many tens of thousands of people migrating for a better life never live to see it. The United Nations has recorded the cases of more than 72,000 migrants who died or disappeared in their journeys from 2014 to 2025, but this is surely a tremendous undercount, given the remote deserts, forbidding jungles, and expanse of oceans that migrants must cross when they have no legal path.[2] There is no telling how many people die every year trying to seek out a better life. The security-first dogma of Western border policies makes these journeys even deadlier than they would be otherwise. As authorities clamp down, migrants are forced to take even more precarious routes to evade detection, putting themselves at increasing risk of dehydration, assault by criminal groups, and shipwreck. The Mediterranean, by far the deadliest migration corridor worldwide, became even deadlier as Italian and EU officials cracked down on lifesaving search-and-rescue operations in the mid-2010s.[3] On the U.S.-Mexico border, the world's deadliest land crossing, aggressive security policies have not necessarily stopped people from crossing, but they have pushed migrants into more dangerous routes deeper into the desert. As the world gets warmer, remote stretches of the desert become deadlier, increasing the risk of dehydration, heat stroke, and exposure.[4]

There are a thousand ways to die in the desert. Juan de León Gutiérrez was a shy sixteen-year-old when he left a small village just down the road from Elena's house in Guatemala's Dry Corridor. It was 2019 and he and a friend traveled together, hoping to find work and a better life than what could be provided by the withering coffee plantations back at home. Two weeks later, he was apprehended by U.S. authorities at the border and sent to a migrant youth shelter. Less than a month after leaving home, he died after undergoing surgery to treat a rare disorder called Pott's puffy tumor, which is not actually a cancerous tumor but a swelling in the forehead typically caused by a sinus infection.[5] He likely acquired the infection during his trip north through Mexico and into the arms of the best-funded border control in the history of the world.[6] He was one of five Guatemalan migrant children to die while in U.S. custody in six months.

What if climate change actually means that fewer people migrate, not more? While climate change can wipe out people's wealth and push them to leave, it can also plunge them deeper into poverty and force them to spend their meager resources just on maintaining their homes and lives, rather than investing in travel. It is a paradox that, despite what one might expect, the countries most affected by climate change experience relatively low levels of emigration.[7] Many don't move because they cannot.

Others might conceivably be able to move but for a thousand different reasons do not want to. In the Pacific, moving tends to mean abandoning entire islands to the seas rather than just a square plot of land, and can involve far more profound trade-offs. Fijians, for instance, have a notion of *vanua*, which refers to the interplay and connections between the natural environment and humans alive and dead.[8] Leaving the place where they were born and raised would be akin to leaving behind their identity, abandoning all that makes them who they are. Similarly in Guatemala, some Indigenous Maya youths fear that abandoning their homeland to head to the United

States will mean the death of their culture, which was already brought to the brink during the thirty-six-year civil war. Going north would also surely mean facing the same racist abuse and indignities that confronted their relatives who migrated in earlier years. "There you will endure more exploitation than here," one migrant explained to his children. "Here is your home, here with your family, here speaking your language, here with your culture, here with your daily life, not there. There, everything changes, absolutely everything, there you can't be free like you are here." [9]

The same dynamic was at work in the United States after Hurricane Katrina. Despite the devastation, many New Orleans locals figured they'd rather rebuild than leave and give up everything they had ever worked for. In the words of an eighty-two-year-old Lower Ninth Ward resident, who was trying to put together the remnants of his home just four blocks from where the levee burst, "I ain't going no place, man. I'm going to stay right here. This is it. This is my home, and this is where I'll be. . . . I ain't about to leave. It took me too long and I worked too hard to build what I had here to just pick up and leave like that." [10] These are the places where people were born, the resting places of their ancestors, and the lands where their entire culture was founded and shaped. Homes are sacred things, not to be tossed aside lightly.

Rarely are the pressures ever straightforward. The difference between choosing to do something, being strongly compelled to do something, and doing so by force can be razor thin. These are complicated questions involving sensitive considerations about people's individual agency and their ability to set their own destiny. If a young adult insists they cannot leave home because they have to care for an aging parent or tend the dying family business, are they staying by choice or by force? It is ambiguous. Certainly the people most ill-suited to claim that one person is staying by choice and another is trapped are armchair academics, researchers, and journalists thousands of miles away. And just because someone's economic outlook would likely improve if they left home does not necessarily mean that it would be a better choice for them. There is a very dark history of forcing marginalized people off their lands, and the simple fact that someone is poor is not proof that they and their family should leave home. But it is clear that there are some people who want to leave their homes and cannot do so, and some who are determined to stay and bear the full consequences.

If she could migrate away from the Dry Corridor, the data suggest that Elena would be financially better off. Migration, by and large, tends to be good for people. Although the actual act of migrating is difficult and expensive, the economic payoff is great. Worldwide, migration has lifted millions of people out of poverty—probably billions. According to the World Bank, migrants going from a lower-income country to a higher-income country typically see their wages grow between three and six times.[11]

These kinds of gains cannot be explained by other factors, such as the likelihood that migrants are by nature simply more entrepreneurial than nonmigrants. Researchers who have looked at immigration lottery systems such as those in the United States and New Zealand—in which the governments use a system of pure chance to distribute a certain number of visas—have compared applicants who were not selected with those who were; the migrants earned multiple times more money than their nonmigrating counterparts within the first year, and the benefits held for the long term. In one study from New Zealand, a decade after arrival, migrants continued to earn roughly triple what they might have earned in their homeland, had better mental health, owned more vehicles, and had more durable assets.[12] Additionally, migrating also had a serious impact on migrants' immediate family members, meaning that there is a sort of trickle-down effect. Truly, these people won the lottery.

Researchers have noted a general hump pattern of who moves and why. If one's country has very low average incomes, migrating abroad is simply too expensive to be feasible. But the odds of migrating increase as incomes rise. Higher incomes also tend to correspond with better education, more urbanization, and people with more cosmopolitan sensibilities, all of which can motivate someone to move abroad to earn more money. But then emigration levels off once per capita GDP starts to reach somewhere around $5,000 per year. Once per capita GDP reaches about $10,000 per year, emigration rates drop because the benefits stop being so pronounced.[13] Why go through the hassle of moving when the opportunities at home are basically just as good? This is one reason why, for instance, there are about 1 million fewer Mexican immigrants in the United States today than there were fifteen years ago; as Mexico got richer (its per capita GDP was just under $14,000 as of 2023), the relative benefits of moving north declined.[14]

There is also a pattern to where people move, and one key element of that is how much money they can expect to earn. Generally speaking, an increase of $2,000 in mean annual wages makes someone 10 percent more likely to choose to migrate there, according to analysis by the World Bank.[15] The same relative logic holds true whether someone is moving internationally or within their old country, from a low-income region to a bustling city. Even refugees, who are leaving not to earn a higher wage but because they could die if they don't, decide where to go in part because of the odds that they can secure a better life, which is more likely in higher-income countries.

All of this is true generally, and all the more so in our new era of climate migration. It is brutally expensive to be poor, and disasters and other climate impacts perniciously claw away at people's livelihoods and meager resources. Lacking a safety net means that one bad storm or a dangerous mudslide can have existential ramifications. Migrating abroad or just to a higher-income city can not only lift oneself out of poverty, but also provide a foundation to help build resilience in one's hometown. The money that migrants send back to friends and loved ones in their origin communities can help build new protections against disaster or make it easier to rebuild afterward.

As I drove across Guatemala, a future in which more people left home didn't actually seem all that bad. The U.S. media tends to portray all of Central America as distressingly poor, with tin roofs, unreliable electricity, and barely there dirt roads. It is anything but. Even far from Guatemala City, our car glided on smooth asphalt past gleaming strip malls that wouldn't have looked out of place in a suburban Phoenix subdivision. Throughout Guatemala, grand two- and three-story houses tower over the road, peeking above the trees and looking oddly out of place behind rough-hewn wood shacks selling pineapples and tortillas. My hotel in Chiquimula featured two separate swimming pools, the water shimmering in the midday sun, perched in a lush sprawling yard where teenagers played soccer and flirted while parents lounged under the pergola. Nearby, an entrepreneur offered tourists paragliding adventures and the chance to go sightseeing in a helicopter. My driver, Conrado, showed me TikTok videos of thrill-seekers screaming with joy in a giant swing at a similar adventure attraction not far away.

Much of the money used to invest in this growth comes from one place: "Remesas," Conrado said simply. Remittances. Money from the United States that migrants send back via Western Union, a mobile phone app, or a range of other services. About three in ten households in northern Central America receive remittances from abroad, typically about $350 per month in Guatemala.[16] That's only about 5 percent of median U.S. household income but can be a life-changing amount of money in Guatemala that can easily cover expenses or provide a significant down payment on climate protections.[17] In Guatemala, more money comes from remittances than from all foreign exports combined.[18]

On the hillside in the tiny village of Barbasco, money from Consuela's son and her other family members abroad was almost certainly crucial for building her new house away from the mountain edge, helping her avoid the collapsing ground. Elsewhere, in Ghana, remittances help farmers invest in irrigation systems and crop rotation.[19] They also help recipients build houses out of concrete rather than mud, so families can withstand landslides and other disasters, and provide access to electricity and telephones that alert them to upcoming disasters and get help when they need it.[20] In hot, coastal Mexico, remittances help residents—particularly poorer residents—purchase air conditioners to stay cool even in the sweltering summer months.[21] In Bangladesh, some recipients say remittances make up half their household income.[22]

In general, remittances are the biggest way that low- and middle-income countries get money from abroad, far outpacing official foreign aid from wealthy Western countries. Worldwide, the $905 billion that migrants (and their children and others) sent in 2024 was a record, and was roughly the equivalent of the entire economy of Switzerland.[23] That isn't even counting the impossible-to-track remittances sent through unofficial channels, such as gifts of television sets and iPhones that migrants bring back to their families, or the cash that people carry as they travel. No one knows how much money crosses borders this way, but it could be just as much as travels through official channels.[24] This money is set to be even more critical now, since the Trump administration shortly after taking office all but killed the U.S. Agency for International Development (USAID), the world's single largest development agency, withholding tens of billions of dollars from disaster relief, public health, economic development, and other efforts worldwide. The shuttering of USAID will have wide-ranging and impossible-to-predict impacts on global development, which will almost surely severely roll back the global fight against poverty, HIV/AIDS, and other horrors. The agency's death may lead to a dramatic uptick in remittances as a replacement. Historically, remittances tend to increase when the money is most needed, such as during economic crises or in the immediate aftermath of natural disasters. To be clear, this is not just a few dollars here and there—it is a transformational amount of money that is shifting hands, mostly from high-income countries to

the developing world, all because some people moved someplace new, got a better-paying job, and want to help out their
friends and family back home. If there were a Robin Hood of
migration, it would be remittances.

In large part because of these remittances, migration has
long been one of the most effective strategies for lifting people
out of poverty—not just migrants themselves, but also their
families and communities in their homeland. There are direct
links between how many international emigrants a country
has and the share of residents living on less than $1 per person
per day: More emigration means less poverty.[25] This money
can help people buy new cell phones, televisions, automobiles,
and so much more. It puts children and siblings and nieces
and nephews through school. But most of it tends to be spent
on the basics—food, health care, and utilities—that help them
get by day-to-day.[26] In a sense, this should not be surprising.
People know what they need, and if you give them money
they will be able to afford those things. Increasingly, a whole
branch of development work has focused on this logic, relying on so-called unconditional cash transfers to allow people
to help themselves, rather than outsourcing decisions to non-
profits in Washington and London. The trend is typified by the
NGO GiveDirectly which, as the name implies, focuses not
on building new water pumps in remote villages or teaching
goat farmers how to code but simply aims to put cash in poor
people's pockets. Evidence shows that the model works. People
don't squander their money on booze and cigarettes; they use
it to reinforce their roof, send kids to school, and improve their
lives in prosaic but meaningful ways.[27]

*   *   *

In this light, migration is not simply a way to escape impending climate disaster but also a strategy to defend against it. The biggest migration problem may be that there's simply not enough of it. Making it easier for people to leave their home can not only help them flee the most dire disasters but also help them earn money to invest in adaptation and resilience strategies. In fact, some economists say governments should actively spend money to encourage people to migrate, at least to urban areas within their own countries, to boost growth.[28] Subsidizing transportation to cities and helping people find jobs or enroll in new training would mitigate the negative impacts of climate change in rural areas, the thinking goes, and help increase the productivity of cities.

Some countries are ahead of the game and have proactively fostered emigration as a national development strategy. Perhaps the clearest case is the Philippines, a country of more than 7,600 islands in the Pacific, which over the last fifty years has sought to serve as labor supply for the world. Since the 1970s, the country has made it government policy to help Filipinos find work overseas, make sure they are not being abused, and enable them to send money back home. The Philippines was perhaps well positioned to become the world's leading labor exporter; colonization by the United States meant that in the early twentieth century Filipinos were considered U.S. nationals (but not citizens) and could travel to and from the mainland freely, while also ingraining English as one of the country's official languages. Now Filipinos account for one-quarter of all nurses in the United States.[29] The government also launched an explicit overseas employment program in the 1970s to help Filipinos leap into the oil-rich Persian Gulf countries as their labor needs dramatically increased.[30] These countries depend on migrant laborers just as much as they do black sticky crude, to the extent that in the United Arab Emirates migrants out-number natives nine to one. Along the way, the Philippine government created a sprawling bureaucracy designed to monitor, manage, and protect Filipinos abroad. It has also successfully invested in finding new niches for Filipinos in the global work-force. Globally, nearly one in every three seafarers is Filipino.[31] More than 10 million Filipinos live abroad and more than 1 million leave to work internationally each year, typically on a temporary basis.[32] In return, the Philippines receives more than $39 billion in remittances, accounting for almost 10 percent of its GDP.[33] There have also been extensive allegations of abuse, trafficking, and poor treatment of Filipinos abroad, not to mention horror stories of murder and sexual abuse. But the Philippine government has effectively doubled down on the strategy in recent years and shown no sign of stopping.

A word of caution is warranted. Migration may be generally beneficial, but it's not clear that it always benefits communities of origin, even when migrants send back large sums of money. For years, researchers tended to think that mass emigration was a bad thing to happen to a country, because it lost its brightest and most entrepreneurial individuals and the ones left behind were those who lacked the know-how or the means to get out. Not only is this brain drain a short-term problem for countries in the Global South, the thinking went, but it is in fact a key reason why they continued to be underdeveloped. In 1981, scholar Joshua Reichert coined the term "the migrant syndrome" to describe the self-perpetuating downward spiral of emigration, in which people who leave make money abroad, deprive their origin community of local investment and laborers, and make others feel as if migration is the only plausible way to move up in the world. His focus was on people from a rural community in Mexico living in the United States, but the lesson applies broadly.[34] "[M]oney earned in the United States has enabled migrants to raise their standard of living to a level that can only be maintained through recurrent migration," he wrote. "Instead of providing a source of capital with which residents could strengthen the local economy, high annual income earned in the United States has merely encouraged more people to migrate, thereby making the town and its inhabitants increasingly dependent on U.S. wage labor as a source of livelihood. . . . Decades of out-migration have fostered the attitude among townspeople that U.S. wage labor is the key to prosperity and success, whereas only poverty and failure mark the lives of those who remain behind." Reading Reichert now, it's easy to think of the Philippines or Guatemala or dozens of other places. Every year, about 1 percent of the local labor force in El Salvador, Guatemala, and Honduras migrates, winnowing the supply of workers, particularly in agriculture.[35] Losing a steady stream of farmworkers has a different effect on an economy than losing all the doctors or tech leaders, but agriculture plays a central role in these countries' economies, so losses there resonate widely.

Indeed, some countries remain extremely dependent on remittances in a way that must surely spell long-term challenges. In the mountain-dominated Central Asian country of Tajikistan, where more than one-tenth of the people go to work in Russia every year (or at least they did before the 2022 invasion of Ukraine), more than one-third of all economic output comes from the money that migrants and other people send back.[36] About one in three working-age Tajik men typically go abroad, and the country's economy is extremely vulnerable to factors it has no control over, including depreciation of the ruble, Russia's varying appetite for cheap workers, and Moscow's decision to become a global pariah by invading Ukraine.[37] Economically speaking, it's hard to say that Tajikistan is much better off in this situation than when it was a part of the Soviet Union decades ago.

In the Pacific, Tonga is so reliant on international remittances that when the power goes out for a few days, as it did when a tsunami hit in early 2022, many people are totally and utterly in the dark, unable to receive wire transfers even when cash is more precious than ever.[38] Four out of five Tongan households receive some money from abroad, typically around $5,000 per year—more or less equivalent to the per capita GDP.[39] In other words, Tongans abroad tend to send back as much money each month as they would generate if they stayed in the country, over and above everything they spend to build their new life wherever they live. In El Salvador, Guatemala, and Honduras, remittances account for somewhere around one-fifth or one-quarter of GDP.[40]

These are not signs of a healthy economy. While people receiving remittances will often use the money to buy food and daily needs, they will also put it toward preparing more young people to emigrate, in a cycle that is sometimes known as the "remittance trap." Aspiring only to leave one's native country prevents young people from staying and working to build something there. Many countries caught in this trap tend to have relatively stagnant economies, meaning that people may have money but are not necessarily creating new things. Meanwhile, local residents might believe their best source of security comes from emigrants abroad, not their elected leaders, and so they come to expect less from the government. National leaders are not pressured to invest in the economy and instead rely on the huge spigot of money gushing in from outside. It's a dynamic similar to the so-called resource curse, or "Dutch disease" (so named after the Netherlands experienced a financial windfall following the discovery of natural gas deposits in the North Sea in the 1960s), which explains how resource-rich countries that receive a huge influx of money often squander it rather than investing in long-term economic growth. Moreover, relying on remittances may also make countries dependent on dollars or euros rather than their own national currency, driving up prices and giving the government less ability to manage the economy. Finally, sending remittances is not free, and a significant percentage of every transaction goes to the bank or money-wiring company that is processing it. Reducing costs to send $200 from about 6.5 percent on average in 2020 to less than 3 percent by 2030 is a major priority of the UN's Sustainable Development Goals, a global wish list that has helped to shape the massive nonprofit-industrial complex.

Some governments have sought to challenge this trend by proactively reaching out to their diaspora, encouraging foreign-trained natives to come back and start new companies, invest in local firms, or otherwise maintain their ties. These are serious issues. But they are also more long-term developmental projects that may mean little to a family being battered by a hurricane or withering in a drought. For these families, an influx of cash from an emigrant abroad could be a lifesaver, but it is only a temporary solution. Money might buy a new house or build a new seawall, but it can only paper over the deeper underlying challenges of a world getting warmer. If crops won't grow or devastating hurricanes recur every year, a few thousand dollars will not make someplace livable. Governments can help, but only if they spend huge amounts of money. Otherwise, the problems will only fester.

If people are going to leave their rural, climate-threatened homes to find work in cities, governments can prepare themselves to take advantage. Rather than simply helping strangers recover from disaster, local governments can allow incoming migrants to build their own futures and strengthen their economies. In no place is this approach clearer than in southwestern Bangladesh's bustling port town of Mongla. The city has become a veritable magnet for climate migrants, in part because its location just outside the Sundarbans mangrove forest makes it accessible to people fleeing cyclones, rising soil salinization, and other hazards. Mongla boasts Bangladesh's second-largest port, and a new manufacturing zone has drawn thousands of workers to dozens of busy factories making bags, clothes, and other items for foreign companies. The sprawling industrial complex is known formally as the Export Processing Zone (locals refer to the place as EPZ, using the English rather than a Bengali acronym, pronouncing the final letter in the British style, *zed*). But it might as well be called a life raft. The formal creation of the EPZ in 1998 and the port's expansion in 2009 sparked what locals describe as an economic miracle in their small, vulnerable corner of the world, injecting money into communities that risked dying off. Rather than the economic uncertainty and sharp elbows of Dhaka, new arrivals can find well-paying jobs that could not only boost their own status but also help transform Bangladesh. A woman could easily earn a relatively good income of a few hundred dollars per month at a factory here. Even residents who don't work in the factories have benefited from the money being spread around. And the city itself has been booming, growing from a population of about 40,000 in 2010 to something like 110,000 today.

Mongla too is threatened by cyclones, rising sea levels, unsuitable drinking water, and the other climate threats that affect this part of the world. In previous years, it flooded twice per day, on pace with the ebb and flow of the tidal river system's schedule. But city leaders have been aggressive in bolstering the town's defenses. They have built a reservoir to keep and preserve freshwater, a water treatment plant to bring new supplies of drinkable water, and a seven-mile-long system of embankments and flood-control gates to prevent inundation. There are trees along the embankment to try to keep the heat down.

Active in this transition is the mayor, Sheikh Abdur Rahman, a slender man with a white beard. When we met in his grand glass-doored office, he wore a sleeveless black Mujib coat made famous by Bangladesh's founding father, Sheikh Mujibur Rahman (the men are not related). People raced in and out while we spoke, intermittently asking Rahman to sign paperwork, weigh in on new developments, and deal with the mounting problems of a city in rapid transition. On one wall, three dozen screens showed footage from security cameras around the city. Elsewhere, plaques and medals adorned the wood-paneled walls. The influx of migrant workers has been accompanied by a new set of challenges, the mayor told me. More residents means more sanitation challenges, pressure on the health care system, and threats of hunger. But growth will continue, he said, and the government is racing to meet the challenge.

Development experts think that Mongla should be a model for the world, or at least other parts of Bangladesh. Experts at the International Centre for Climate Change and Development (ICCCAD), a Dhaka think tank, have plans to replicate Mongla's success in more than two dozen similar towns across the country. "Mongla has the livelihood opportunities in export, EPZ, the port, and also the shrimp farming, because there is a local fish market, and tourism business," said Lutfor Rahman, an ICCCAD researcher I met at his organization's expansive ninth-floor offices in a university building. "The people who are migrating want to migrate [because of] failure to adapt. . . . It could be the natural disaster like river erosion, floods, or cyclone, they come to the nearby city. This is the concept."

Only time will tell if Mongla's model inspires copycats. While governments have invested huge sums to try to maximize the benefits of remittances, far fewer destination communities have sought to capitalize on an influx of climate migrants. They might want to consider it. If people are already moving, why not use the situation to everyone's advantage?

7

Outrunning Bullets and Disasters

Under the right circumstances, graffiti can be a remarkably brave thing to write. "The people want the fall of the regime," a group of boys spray-painted on a wall in Syria. It was March 2011, and the phrase had become a calling card of the blossoming Arab Spring, a rapidly growing movement for government accountability and against corruption that swept across North Africa and the Middle East. The phrase was born three months earlier and roughly two thousand miles away, when a fruit vendor named Mohamed Bouazizi lit himself on fire outside a provincial government building in a small town in central Tunisia, in an extreme act of desperation and protest. By March, the presidents of Tunisia and Egypt had stepped down, and in Syria protesters were demanding that dictator Bashar al-Assad, a UK-trained ophthalmologist, follow suit. "It's your turn, doctor," the boys wrote on their school wall in the southern city of Daraa. Several added their own names, as if to add a signature to their artwork.[1]

The government responded with an iron fist. At least sixteen boys accused of writing the graffiti were arrested and tortured, triggering outrage that engulfed the nation. As weeks turned to months and months to years, the protest evolved into an uprising, which grew into a rebellion and became a civil war and then a proxy conflict for major powers across the world, as Western forces supported some rebel factions, Iran and Russia bolstered Assad's regime, and the vacuum in between fostered the growth of the extremist Islamic State. For nearly fifteen years the government held on, until a surprise, lightning-quick rebel assault in late 2024 seized power virtually overnight while Assad's allies were preoccupied elsewhere. Assad fled and left ostensible control of the country to a former jihadist named Ahmed al-Shara. Over the course of the war, hundreds of thousands of civilians were killed and roughly half of Syria's population was displaced.[2]

Few twenty-first-century conflicts have shaped the current world as profoundly as the civil war in Syria. Like all conflicts, Syria's ghastly war can be traced to a multitude of factors. Millions of people choose violence for reasons all their own, which are often hard to pin down. The broadening Arab Spring was one driver, offering long-oppressed people a glimmer of hope that a different system was possible. So too was the government's continued brutality and refusal to cede to the people's demands. Assad, like his father, Hafez al-Assad, had created a tinderbox that was just waiting to ignite. There were also sectarian divisions within Syria, not to mention that, when the war broke out, the country was sheltering 1 million Iraqi refugees who had fled in the wake of the U.S.-led invasion of Iraq and who were struggling to build a new life.

Another factor was climate change. Starting in 2007 and lasting at least until 2010, Syria underwent the worst drought in its modern history. Tens of thousands of farmers lost all or most of their livestock. The country's northeastern breadbasket dried up, and crops were affected by a historically pernicious outbreak of the fungal disease wheat rust, which can also be traced to climate change. Poor farmers had virtually nothing to show for their harvests. Agricultural self-sufficiency had been a source of national pride, but for the first time in a decade the country needed to import large amounts of wheat.[3] Assad's regime had also cut food and fuel subsidies and declined to change course amid the drought, aggravating the situation.

In response, as many as 1.5 million people left their farms to move to cities. They went to slums and shantytowns, borrowed money where they could, and tried to restart their lives in the face of crime and corruption. They also competed for jobs and scarce public services with the newly arrived Iraqi refugees. In 2010, a year before war broke out, about one-fifth of all residents of Syria's cities were either refugees or internally displaced people from elsewhere in the country.[4] Syria's overall population was growing at a rapid rate around that time, especially in and around the cities, which were fast becoming hotbeds of frustration and desperation. In just about a decade, the population of towns outside cities like Aleppo ballooned from around 2,000 to nearly 400,000.[5] As in countries worldwide, cities make big promises that they often cannot fulfill, leaving many migrants disappointed and poor in a brand-new way. The result, according to researchers, was anger, particularly at the government. When the protests started, more people were ready to join the chorus of opposition.[6] The drought-prompted migration had a "catalytic effect" on the conflict, concluded a team of researchers led by the climate scientist Colin Kelley, ramping up the underlying tensions and making violence more likely.

Of course, droughts have hit the Middle East for all of history. But human-caused climate change was causing annual surface temperatures in the region to rise faster than global averages and made it two or three times more likely that there would be a drought this severe, Kelley and his team concluded.[7]

There are any number of ways in which a responsive, well-functioning government could have stepped in to ameliorate the situation, but the government failed to meaningfully respond to these climatic shocks in a way that might have helped people. Worse, years of corruption and mismanagement had seemingly laid the groundwork for the drought to be particularly pernicious.[8] "When the drought happened, we could handle it for two years, and then we said, 'It's enough,'" a thirty-eight-year-old former farmer named Faten told the *New York Times* in 2013. "So we decided to move to the city. I got a government job as a nurse, and my husband opened a shop. It was hard. The majority of people left the village and went to the city to find jobs, anything to make a living to eat.... Since the first cry of 'Allahu akbar,' we all joined the revolution. Right away.... Of course, the drought and unemployment were important in pushing people toward revolution."[9]

Increasingly, climate change is both magnifying the underlying drivers of conflict and striking the people fleeing war particularly hard. Rarely is the fighting as extreme as in Syria, and climate change has not yet been a primary driver of a war between countries, although it could be someday. Climate-linked violence tends to be relatively low-level skirmishes between communities, militias, and local authorities. But history has repeatedly shown that if a government fails to respond to simmering tensions and low-level conflict, the challenges can escalate. In certain situations, researchers have been able to quantify how specific increases in temperature and changes in precipitation are systematically correlated with greater likelihood of conflict. A meta-analysis of fifty-five studies found that one standard deviation increase in temperature increased the odds of conflict between groups by 11 percent and the odds of interpersonal conflict by 2 percent.[10] "If future responses to climate are similar to these past responses, then anthropogenic climate change has the potential to substantially increase global violent crime, civil conflict, and political instability, relative to a world without climate change," conclude the authors Marshall Burke, Solomon M. Hsiang, and Edward Miguel.

The chance of climate-induced violence is acute for communities who depend on agriculture, since climate change represents a radical economic threat. In some cases, the impact is direct: Climate change shrinks the availability of valuable resources such as water, leading fights to break out. More often, however, the process is indirect, and migration can easily form the connective tissue that causes climate change to lead to deadly conflict. A common theme is that migrants create the perception of a threat to an underlying social, economic, or political order that was already fragile because of dwindling resources. When climate change pushes too many people together into a confined, resource-strapped space, deeper societal fault lines can become exposed, authorities' weaknesses can become magnified, and long-simmering tensions can start to boil over. As economic times get tough, people may solidify allegiances to their ethnic or religious group, which can deepen intergroup divides. Often, problems do not even need to actually occur, but simply the perception that migrants might undermine economic or political systems can provoke conflict.[11] New arrivals are also easy scapegoats for economic challenges and tend to take the blame for deeper, harder-to-solve issues around housing, jobs, and demographic change.[12] If people from climate-vulnerable areas are forced to move, President

Barack Obama said in Leonardo DiCaprio's 2016 documentary *Before the Flood*, "then you start seeing scarce resources that are subject to competition between populations. This is the reason the Pentagon has said this is a national security issue—this isn't just an environmental issue; this is a national security issue." [13]

Conflict can also erupt in the places people are fleeing. The ones who leave are typically community leaders and entrepreneurs, while those left behind tend to be the poorest and the most desperate. As a dying village plunges further into poverty, conflict-resolution mechanisms break down and violence can be quick to flare up. [14] There are always other issues at play, and local conditions are ultimately what determine if people's unease will turn into protest and then become violent. When the actual shooting starts, the cause is almost always a combination of these and many other factors happening in conjunction.

Political scientists generally agree that there are some more or less universal conditions that can tip a situation of domestic unrest into violent conflict. One of them is a large underclass of unemployed or underemployed men and boys who could easily be transformed into fighters. Other factors are gross inequality between the rich and poor, a rapid change in demographics, and a stagnant economy. [15] These are precisely the phenomena that can occur when natural disasters destroy people's homes or climate change undermines their chance to earn a living, forcing them to migrate to look for work.

Yet the climate-conflict link is not limited to the Global South. In the United States, researchers have repeatedly identified that environmental factors such as heat can worsen hostility. In Baltimore, Chicago, and elsewhere, gun violence is empirically more likely to occur on warmer days.[16] In prisons without air-conditioning, violence is 18 percent more likely on hot days.[17] Generally speaking, the risk of violence increases with the temperature.[18]

We must be careful here. While climate change can aggravate underlying drivers of conflict and force people into new areas, violence is not inevitable, nor does it mean that people displaced by climate change are themselves security risks. Climate change on its own did not cause the Syrian revolution; rather, it was a proximate factor that exacerbated peoples' grievances against the government, pointed out the regime's years of policy failures, and illustrated leaders' poor governance. The drought and the migration it inspired were a test of the Assad regime's economic development policies and its capacity to respond to citizens in need; the regime failed spectacularly.

The test is coming to other parts of the world, in few places more obviously than Africa's Lake Chad region. Sometimes referred to as "the world's most complex humanitarian disaster," the region is also a vibrant home to around 100 million people, with a lake shaped like the outline of a swan, with a squat body perched at the border of Cameroon and Chad, a neck snaking up into Nigeria, and an overgrown head peeking into Niger.[19] It's a place that rarely makes Western news for positive reasons. On the few occasions that Lake Chad finds its way into a newspaper headline, it's often connected to actions of the extremist Islamist group Boko Haram or its recent successor, the Islamic State West Africa Province. The group has clashed with regional forces for years, particularly those of Nigeria, and sought to overthrow the government

and establish an Islamic state. In 2014, Boko Haram's kid-
napping of hundreds of schoolgirls from the town of Chibok
captured international attention and spurned an urgent plea
to "#BringBackOurGirls" from First Lady Michelle Obama
and many others.[20] That year, it was considered the world's
single deadliest terrorist organization, responsible for a star-
tling 6,644 deaths, many of them targeting individuals and
organizations it deemed to be aligned with Nigeria's secular
government or the West ("Boko Haram" translates to "Western education is forbidden" in the local Hausa language).[21]
An estimated fifty thousand people were killed in the conflict
from 2002 to 2022, and more than 2.5 million displaced.[22]
At the same time, escalating conflicts between seminomadic
Fulani herders and more sedentary farmers have repeatedly
spilled into low-level clashes, causing more than fifteen thousand deaths across the region from 2010 to 2021, primarily
after 2018.[23] This violence lacks the spectacle that would
earn it a White House media campaign, but it has nonetheless
contributed to destabilizing a region with one of the world's
fastest-growing populations.

Here, the history of migration, environmental strain, and
conflict are deeply interconnected. For many people around
Lake Chad, movement during different times of the year has
long been a way of life. But repeated droughts and a booming population have placed excess strain on limited resources,
pushing people to move and, like in Syria, aggravating a suite
of grievances against the authorities. Most symbolically, Lake
Chad itself has shrunk dramatically, splitting into two separate
pools. Once the planet's sixth-largest inland body of water,
its surface area shrank nearly 90 percent, from about the size
of Massachusetts in the 1960s to smaller than Jacksonville,

Florida, in the 1990s, although it's been relatively stable since then. Meanwhile the lakeside population grew precipitously, from about 700,000 people in 1976 to 2.2 million in 2019. Across the greater Sahel region—the semiarid four-thousand-mile belt separating the Sahara from the tropical savannas—temperatures are rising one and a half times faster than the global average.[24] As water and other resources have become more scarce, more people have abandoned their traditional livelihoods and moved to cities. Like in Syria, migrants may themselves be in poor or worsening socioeconomic positions, without work, and adding to the pressures on local governments to provide services and supports. When governments fail to adequately respond, they undermine their own legitimacy. The economic effects extend upstream to traders, merchants, and craftspeople who depend on now-inaccessible downstream markets.[25] People start looking for new sources of authority and, due to history and demography, often embrace their ethnic and religious identities in ways that can make local jihadis seem like natural allies. Boko Haram and other Islamists have also attracted supporters by offering material benefits in the form of welfare systems, microfinancing, and marriage arrangements.[26] They have dug wells, distributed seeds, and offered cash, which can be especially attractive to struggling herders and farmers.[27]

One group of players in this process is the millions of *almajirai*: boys and young men often from poor rural families who are sent to the city and handed over to Muslim leaders for religious education. Their living conditions tend to be meager and lonely, their training limited strictly to the Islamic faith, leaving them mostly unequipped to engage in profitable work after finishing school. *Almajirai* is a Hausa word that derives from

the word for migrants but has come to be synonymous with local beggars due to the long hours of panhandling that students
perform.[28] The centrality of the *almajirai* to Boko Haram is a
topic of debate, but they have been associated with regional Islamist movements for decades; in 2012, Nobel Prize–winning
writer Wole Soyinka called them the "foot soldiers" of northern Nigerian militias, including Boko Haram, who had been
forced into their position by economic stress and are easily
manipulated by populists and militant leaders.[29]

Jihadist groups have both benefited from and inflamed the
tensions between regional farmers and herders, who have repeatedly found themselves competing over winnowing or unpredictable resources. At its worst, this kind of conflict can lead
to deadly massacres between communities, such as one inci-
dent in 2023 in which gunmen descended on several remote
farming villages in Nigeria's Plateau State, killing more than
one hundred people.[30] The attack was reportedly precipitated
by cattle destroying a banana plantation; the plantation owner
complained, and in retaliation Fulani herders descended on the
community with impunity.[31] Authorities tend to be slow to respond to these kinds of killings, leaving both sides simmering
with resentment at the state and making it more likely for them
to align with militants who represent some sort of authority. At
the same time, increasing fears about terrorism have led countries to ramp up border security and make it harder for herders
to migrate in times of drought, as they have done in years past.[32]

While people displaced by climate change can add to the pressures that create conflict, violence in places like the Sahel or
Syria also causes massive displacement that puts huge numbers

of people at new risk of climate disaster. Three-quarters of the world's forcibly displaced people live in countries that are heavily impacted by climate change, a reflection of the fact that the vast majority of refugees and other humanitarian migrants live in or next to their conflict-battered home countries.[33] Even within these countries, refugee settlements tend to be located in remote places and are erected hastily, often with weak materials, all of which makes them vulnerable to severe weather. By 2050, most refugee settlements will experience twice as many days of dangerous heat as they did in 2024.[34] Often, conflict and climate have combined, creating a tailspin of crises that amplify one another; people may try to outrun the bullets, but sometimes disaster is harder to escape.

Such is the case in southeastern Bangladesh, where since 2016 about 1 million ethnic Rohingya people have fled from a genocide targeting them in neighboring Myanmar. UN investigators have concluded that Myanmar's military regime has committed crimes against humanity against the Rohingya, including the torture of children, rape, and cold-blooded executions.[35] Even though some Rohingya can trace their families back generations, many people in Myanmar insist they have entered the country illegally from Bangladesh. Many won't call them Rohingya, instead using "Bengali" or more pejorative names.

It was here, on a thin stretch of land where so many Rohingya have fled for safety, that I met Zara, a fifty-year-old woman with a hard smile. She told me that the military came to her village, known alternately as Kyar Gaung Taung and Rabillya, and raped and killed her neighbors in cold blood. Old, young, it did not matter, she said. The village was effectively destroyed; government troops arrived in the middle of the night and began firing indiscriminately. Those who fled were shot at with rifles and RPGs from helicopters. Some of Zara's neighbors—including a woman who was eight months pregnant and an eighty-five-year-old man—were taken captive and tortured by the Burmese military, beaten and forced to sit in so-called stress positions with their hands on their head and their eyes on the ground, in the sun without food or water, for up to eight hours.[36] Zara told me she believed one of her sons was killed in the fighting, though she never saw his body to confirm it. Her other son was sitting in a jail cell in Malaysia, convicted of human smuggling. Thousands of Rohingya have paid traffickers to shuttle them from Myanmar to Malaysia, Thailand, and other countries, with the hope of starting a new life.

But when I met her, Zara had other matters on her mind. She had been awake all the previous night, as a deadly cyclone barreled through the Bay of Bengal and tore at her nearby hut, built from posts of bamboo, walls of black plastic sheeting, and a thatch roof. At least six people were killed in the storm, as winds of seventy-five miles per hour destroyed tens of thousands of shelters.[37] Overnight, one of the plastic seams in Zara's wall ripped, letting water spill into her home. As I trudged through ankle-deep water, she noted how the structure's central bamboo pillar had cracked but not broken in the winds. It was afternoon by the time I met her, and a heavy branch had been wedged into her front doorway to prop up the ceiling. A new weight was added to hold the roof down, in case the winds returned. The small home had survived intact, if just barely.

Others were less lucky. Just a few steps away, across a low-lying swamp that used to be a field, lay a tangled pile of grass, vines, and debris that the previous day had been the home of Abul, a local imam, and his four family members. The hut collapsed in the night when heavy winds sent tree branches flying. The family of five crawled out through what had once been the door, sliding through inch-deep mud as neighbors pulled them to safety. Now a cow trod over what used to be the roof and was idly chewing on the detritus of their home. Nearby, an uprooted tree, fifteen feet tall, was lying on its side. A bamboo outhouse built by a foreign aid group was in tatters. The road in front of the village, which runs down from the beach town of Cox's Bazar, was covered with fallen branches. A soft drizzle was falling. There was a stiff breeze.

Zara had arrived from Myanmar just three months previously. Like many Rohingya, she and her new neighbors were not in an official UN refugee camp but in a makeshift community outside. While UN officials made efforts to move official camp residents into schools and other solid structures ahead of the cyclone, that wasn't an option for these people outside of their aegis. Per Bangladesh government policy, the United Nations did not provide protection to these informal communities, and many residents there felt abandoned. "They are probably extra vulnerable to anything like this," Shinji Kubo, the country representative for the UN Refugee Agency, told me. In Bangladesh and fifteen other climate-vulnerable countries, 40 percent of refugee settlements face a significant risk of flooding.[38] For Zara, the overnight cyclone was yet another terror that left her panicked. She did not feel safe until the sun finally rose in the morning, she told me. But there was no time to rest. Once again, she had to rebuild.

As she did, she received little help from native Bangladeshis living nearby. Bangladeshis have often eyed Rohingya refugees skeptically, in part because the new arrivals have cut down vast forests for firewood, sometimes trekking for up to sixteen hours to gather wood.[39] Tensions between refugees and host communities over deforestation and other environmental issues are not uncommon, said Andrew Harper, a special advisor on climate action to the UN Refugee Agency, UNHCR. "Not only are refugees coming from countries which are impacted by climate change, this climate change—it's not focused on a country; it's regional. It's often the host states that are also impacted as well," he told me.[40] "So they're seeing a surge in the population on their territory, and this population comes with no resources, or generally very little in the way of resources and assets." International organizations like his need to step in to help local communities, he said, to prevent tensions from escalating into conflict. "If you don't protect the environment in which displaced people are, you cannot expect that host community to protect the refugees." In Bangladesh, the influx of Rohingya prompted a series of reforestation initiatives.[41]

The government has also shuttled tens of thousands of refugees to an island more than a hundred miles away, called Bhasan Char. While ostensibly part of a plan to reduce overcrowding in and around the refugee camps and offer refugees an opportunity to start anew, the policy has come under fire. The island is extremely vulnerable to cyclones, and refugees are unable to come and go from it freely. The UNHCR has some oversight of the refugees' conditions, but critics have all the same warned that it could become "a UN-supported prison island."[42] The relocation illustrates a difficult tension at the heart of the international protection system. Most people forced to flee end up not far from where they started and often find safety only because of the generosity and solidarity of their neighbors. Yet these neighbors tend to be only marginally more secure than the refugees themselves, and when weeks of displacement turn to years—especially as impacts of climate change grow worse—fatigue sets in. International support tends to fade as crises stretch on and the challenges morph from those associated with erecting tents and latrines into more complex ones involving opportunities for children to go to school and for adults to find work. How is a poor country expected to accommodate a million desperate people for a decade?

8

Population Panic and the Environmentalist Seeds of the Modern Anti-Immigration Movement

When leaders in Washington, Brussels, and other Western capitals look at the climate migration puzzle, their response is typically neither to cut their greenhouse gas emissions nor to ensure people can move safely and orderly. Instead, they pull up the drawbridge. That is especially the case for places like Syria and the Sahel, where security officials see climate-linked violence not as a tragic result of the West's emissions but as a security threat. A new generation of overseas criminals, warlords, and gangsters is being born in the most climate-vulnerable places, they warn, desperate to infiltrate the Global North. "As families risk their lives in search of safety and security, mass migration leaves them vulnerable to exploitation and radicalization, all of which undermine stability," U.S. Defense Secretary Lloyd Austin said in 2021, making the connection explicit.[1] If refugees in general are often viewed suspiciously, the same is true of climate migrants.

For more than a century, this conflation of immigration and environmental threats has repeatedly manifested in immigration restrictions under the guise of protecting the country's resources and preventing despoilment from outsiders. The narrative has had particular salience in the United States, which is home to more immigrants than any other country and also is deeply invested in the notion of its own self-sufficiency and ability to wall off the outside world. The journalist and author Robert D. Kaplan gave voice to the thinking in an influential 1994 article in *The Atlantic*, in which he warned that "disease, overpopulation, unprovoked crime, scarcity of resources, refugee migrations, the increasing erosion of nation-states and international borders, and the empowerment of private armies, security firms, and international drug cartels" were paving the way for "the coming anarchy."[2] He focused his argument on West Africa, which he claimed "is becoming the symbol of worldwide demographic, environmental, and societal stress, in which criminal anarchy emerges as the real 'strategic' danger." The region, he wrote, "consists now of a series of coastal trading posts, such as Freetown and Conakry, and an interior that, owing to violence, volatility, and disease, is again becoming, as Graham Greene once observed, 'blank' and 'unexplored.' . . . West Africa's future, eventually, will also be that of most of the rest of the world."

In reality, West Africa is a flourishing region with an exploding economy and vibrant cultural and tech scene. While the challenges of climate change, corruption, terrorism, and other ailments are real, they are by no means the defining features of a place that hundreds of millions of people call home. All the same, fears like Kaplan's have continued to grow, positioning climate change as a driver of global poverty and conflict that is desperately clawing at the door of the pure and unspoiled (and typically implicitly white) West. Security contractors, politicians, and the military-industrial complex have increasingly latched on to this narrative to secure high-dollar contracts to enforce the border and detain immigrants, while simultaneously advancing both their careers and a storyline in which the

Global North is somehow the victim of global climate change, not its perpetrator. Even the climate-denying Trump administration has been willing to embrace the logic. "Climate change is impacting stability in areas of the world where our troops are operating today," Trump's first Pentagon chief, James Mattis, told the Senate in 2017. "It is appropriate for the combatant commands to incorporate drivers of instability that impact the security environment in their areas into their planning." [3]

This logic has a long history, much of which can be traced to English philosopher and economist Thomas Malthus, whose 1798 *An Essay on the Principle of Population* outlined a tragic death spiral that can occur when the population swells to unsustainable levels.[4] An excess of people will quickly consume all the society's resources, Malthus argued, causing it to burst at the seams and undo itself by starvation. To avoid this dire fate, Malthus proposed two solutions: a "preventive check" that discouraged people from having children, such as by making the cost of rearing a large family prohibitively expensive, and a "positive check" to allow people predominantly from "the lowest orders of society" to suffer disease and malnourishment without intervention such as government welfare. This second check was perhaps most radically illustrated by Ebenezer Scrooge, the villain of Charles Dickens's *A Christmas Carol* (itself a thinly veiled attack on Malthusianism). Early in the story, Scrooge complains about poor and "idle" people who should go to prison or poorhouses rather than depend on charity. "If they would rather die," Scrooge says, "they had better do it, and decrease the surplus population." [5]

Malthus's argument seems hardly relevant for our modern,

postindustrial world. The population of his native United Kingdom is now six times larger than in Malthus's time and yet it has easily avoided pestilence and famine, all the while adding a social safety net that, while sometimes maligned, is leaps and bounds more robust than anything imaginable in his Georgian era. Yet Malthus's narrative persisted, particularly among early environmentalists who found in his work an argument to protect natural resources. In the process, it morphed into inspiration for generations of anti-immigrant zealots. Frequently, this has overlapped with explicit racism. For instance, President Theodore Roosevelt, who created five national parks and 150 national forests, decried "race suicide" by white people who declined to have children and were being replaced by immigrants and ethnic minorities.[6] It was a sentiment that was on the rise at the time, alongside the growing eugenics movement. John Muir, a co-founder of the Sierra Club and an icon of environmentalism, was also a racist who referred to Indigenous people as "savages." David Starr Jordan, another early Sierra Club leader, was an explicit white supremacist who advocated for forced sterilization of people he deemed eugenically unfit. As recently as 1989, the Sierra Club's board stated that U.S. immigration "should be no greater than that which will permit achievement of population stabilization in the United States," although it has since vocally backed away from this position.[7]

This tension is still visible in the current conservationist movement, although it is thankfully significantly reduced. The bridge between then and now can be traced to a single book that arrived just as the U.S. government was beginning to take a more active role in protecting the environment. In 1968, *The Population Bomb*, by biologist Paul Ehrlich and his uncredited conservationist wife, Anne, was a landmark document that predicted imminent global panic, famine, and catastrophe due to overpopulation. It provided perhaps the most stirring argument about the dangers of resource scarcity since Malthus more than a century and a half earlier, and it gave birth to a new era of population panic.

"The battle to feed all of humanity is over," Ehrlich claimed on the book's very first page.[8] "In the 1970s, the world will undergo famines—hundreds of millions of people are going to starve to death in spite of any crash programs embarked on now." Huge swaths of the planet, he said, were doomed and could not be saved. All the same, "immediate action" to control global population growth could help future generations avoid suffering the same fate: "We must have population control at home, hopefully through a system of incentives and penalties, but by compulsion if voluntary methods fail. We must use our political power to push other countries into programs which combine agricultural development and population control. And while this is being done we must take action to reverse the deterioration of our environment before population pressure permanently ruins our planet." Two pages later, Ehrlich describes the claustrophobia he felt riding a taxi through a crowded slum in Delhi: "The streets seemed alive with people. People eating, people washing, people sleeping. People visiting, arguing, and screaming. People thrusting their hands through the taxi window, begging. People defecating and urinating. People clinging to buses. People herding animals. People, people, people, people. . . . Since that night, I've known the *feel* of overpopulation."

The Population Bomb was one of those rare books to emerge from the stodgy confines of academia and make a splash in popular culture. It was a bestseller, with more than 2 million copies sold and translations in multiple languages. Paul Ehrlich also became a cultural figure on par with, in more recent decades, the astrophysicist Neil deGrasse Tyson, economist Paul Krugman, or philosopher Bernard-Henri Lévy. He and his dark, heavy eyebrows became a regular guest on *The Tonight Show Starring Johnny Carson* and a clear favorite of the comedian host. One of Ehrlich's first appearances, in April 1970, "elicited probably more mail than any guest at that time that we have had on the show," Carson said a decade later, during one of Ehrlich's many subsequent appearances.[9] Ehrlich's deep, steady voice suggested confidence and authority, reminiscent of a just-the-facts evening newscaster.

The popularity of *The Population Bomb* was due in large part to its sweeping and alarmist claims about imminent devastation. Needless to say, the predictions were at best hyperbole and at worst downright racist fearmongering. Although the globe's population has more than doubled since the book's release, no such population bomb has detonated. Around 3 million people died as result of famine in the 1970s, mostly in Asia. These lost lives were tragic—and overwhelmingly preventable. They were not the result of overpopulation. Famine's death toll in the 1970s was a fraction of its roughly 13 million victims in the 1960s, and far less than the hundreds of millions Ehrlich promised.[10] The decades since then have seen a remarkable reduction in the number of people dying from famine, largely because of the so-called Green Revolution in agriculture, which allowed for farmers around the world to use new pesticides and new varieties of high-yield and durable crops, alongside modern irrigation and other techniques that had previously been used only on a limited basis by farmers in wealthy countries. This Green Revolution was aided in part by an increasingly sophisticated international humanitarian sector that arose in the aftermath of World War II, including elements of the United Nations, as well as the emergence of global networks that led to cheaper food and massive reductions in poverty. Since then, famine has largely occurred during war (when aid groups and global supply systems are unable to assist starving people) or as a result of clear political and policy choices, usually by totalitarian regimes. One prominent example is North Korea's so-called Arduous March from 1994 to 1998, during which somewhere between several hundred thousand and 3.5 million people died because of a combination of poor economic policies, Russia's withdrawal of aid following the collapse of the Soviet Union, and crop failures. These people did not starve because North Korea's population was too large; they died because complicated global trends and domestic policy were wielded against them. Similarly, more recent famines in Yemen, South Sudan, Ethiopia, and Gaza were the result of wars and blockades that have prevented imports of food. The problem is not that there was not enough food but that specific militaries kept it out.

In 2009 the Ehrlichs backpedaled a bit on their alarmism. They called the opening lines of *The Population Bomb* "troublesome" and expressed regret about some of the rhetoric they had used.[11] They acknowledged that the book failed to anticipate the massive impact of the Green Revolution on the global food supply. Yet they nonetheless doubled down on their central point, that overpopulation remains a serious problem for the planet and its ultimately finite resources. "Signs of potential collapse, environmental and political, seem to be growing," they claimed in an academic article reflecting on the book's legacy. "The pattern is classic—population grows to the limits of current technologies to support it, followed by technological innovation (e.g., long canals in Mesopotamia, green revolution in India, biofuels in Brazil and United States) accompanied by more population growth and environmental deterioration, while politicians and elites fail to recognize the basic situation and focus on expanding their own wealth and power."

They failed to confront a deep underlying tension in their book, however, which is that population control measures tend primarily to target poor, brown, and Black people from former colonies in the Global South rather than wealthy white people from Europe and North America. Despite the vivid description of Delhi as a preview of humanity's overpopulated future, the city's population in 1968 was a fraction the size of Paris's or New York's.[12] Somehow, *The Population Bomb* didn't seem as repulsed by crowds in Times Square or along the Champs-Élysées. Historically, meanwhile, the major culprits of greenhouse gas emissions and other causes of environmental devastation have been wealthy countries in Europe and North America.

To be clear, *The Population Bomb* was far from the only voice claiming global population was out of control and needed to be addressed. At the time, the globe was about two decades out from World War II, which devastated parts of the world, and people were coming to grips with the new world order. For one, many countries were seeing a dramatic baby boom and public health measures that extended life expectancy; meanwhile, there was a lull in outright war between countries. The global population was indeed growing at a remarkable rate, and it was not absurd to be anxious about where that might lead. There were around 1 billion people in the world in 1800 and something like 1.7 billion in 1900. That number hit 2.5 billion by 1950, and nearly another billion arrived over the next fifteen years.[13] By 2022, the global population reached 8 billion.

Population growth was top of mind for leaders around the globe. UN Secretary General U Thant warned in 1969 that "the population explosion" posed as dire a threat to humankind as the Cold War arms race, environmental destruction, and global poverty. That same year, just after the release of *The Population Bomb*, the United Nations founded the UN Population Fund (known as UNFPA for its pre-1987 name, the UN Fund for Population Activities). In 1972, an informal group of experts and businesspeople called the Club of Rome published *The Limits to Growth*, a report summarizing a mathematical model that predicted "the limits to growth on this planet will be reached sometime within the next one hundred years. The most probable result will be a rather sudden and uncontrollable decline in both population and industrial capacity."[14] Reading through *The Limits to Growth* now, I couldn't help but think of Hari Seldon's use of "psychohistory" to predict the end of the galactic empire in Isaac Asimov's sci-fi classic Foundation series. "Psychohistory was the quintessence of sociology; it was the science of human behavior reduced to mathematical equations," Asimov describes. "The individual human being is unpredictable, but the reactions of human mobs, Seldon found, could be treated statistically."[15] Except that in the fictional series, at least, Seldon's predictions actually came to pass. The same cannot be said for population alarmists.

*   *   *

In response to these fears, one of the so-called checks proposed by Malthus seemed to resonate widely: prevent people from having children, by force if necessary. One manifestation of this urge was increased global support for family planning. This cut both ways: Organizations such as UNFPA have done tremendous work to make birth control more accessible and allow more women to decide when, if, and how they want to have children. But they have also contributed to gruesome postcolonial campaigns to reduce births in developing countries. In particular, this kind of thinking undergirded international support for sterilization policies such as India's, which was strongly supported by the World Bank and the United Nations. India's program tended to target the poor and underprivileged. In some cases, police all but dragged poor men to undergo vasectomies. Prime Minister Indira Gandhi in 1975 declared a twenty-one-month period of national emergency and used sweeping powers to force millions to undergo surgery in what has since been deemed one of the gravest civil rights abuses in the country's history. The program peaked in 1977, when 8.1 million people were sterilized, many through coercive methods.[16] Western leaders cheered along throughout. Between 1972 and 1980, the World Bank gave India $66 million in loans expressly for population control purposes.[17] "At long last, India is moving to effectively address its population problem," Robert McNamara, the World Bank president and former U.S. defense secretary, wrote in praise of the emergency.[18] Although the period of emergency is now widely condemned by international critics, India continues to implement family planning and sterilization that tend to disproportionately affect the poor, rural, and lower castes. In 2014, fifteen women died and dozens more were hospitalized at a pair of state-run sterilization facilities in the central Indian state of Chhattisgarh. Meanwhile, China's one-child policy, adopted from 1980 to 2015, led to a wave of forced and coerced methods including abortions and sterilizations. Now China's leaders are reckoning with the exact opposite problem: how to prevent their country's population from shrinking, as it has done since 2022.

Indeed, population panic fell out of fashion in the years since *The Population Bomb* was published. In the 1980s, a new wave of free-market adherents maintained that economies would simply adjust to larger populations, and families confronted by dwindling resources would make the rational choice to stop procreating. Advocates for limiting population growth were part of the "doomsday crowd," who failed to believe in the boundless innovation and inherent optimism of the American economy, insisted Ronald Reagan.[19] UK Prime Minister Margaret Thatcher similarly asserted that "the modern world" was built through abundant innovation and scientific progress. "This is a world which is able to sustain far more people with a decent standard of life than Malthus and even thinkers of a few decades ago would have believed possible," she claimed in 1988.[20]

To Reagan, Thatcher, and their ilk, Malthus and his modern adherents were enemies of free choice and the progress of Western science. Meanwhile, the burgeoning anti-abortion movement wholeheartedly rejected the notion that governments and the United Nations ought to be encouraging abortions and other family planning methods. If anything, new political movements encouraged more children, not fewer.

Cold War politics, in other words, had flipped: people who supported limits on population growth were now seen as adherents of authoritarianism, while laissez-faire policies were all the rage. In the last few years, the alarmists have completed their 180-degree turn. What happens, some worry now, when the number of people globally starts to fall? Among the most prominent voices concerned about population shrinkage is Elon Musk, the multibillionaire and close ally of Donald Trump. "Population collapse is the biggest threat to civilization," he posted on X in 2022.[21] Musk has donated millions of dollars to fertility and population research and has produced more than a dozen children of his own.[22] As with the panic about overpopulation a generation ago, it seems wise to approach this new hysteria skeptically.

In the United States and other high-income countries, concerns about overpopulation splintered into efforts to limit immigration, particularly of people from poorer countries. The direct line from *The Population Bomb* to the modern American anti-immigration movement traces to Zero Population Growth, a group Paul Ehrlich co-founded in 1968 to advance his efforts. Now known as Population Connection following a 2002 rebrand, the nonprofit group has more than $29 million in assets and boasts forty thousand dues-paying members.[23] It does public outreach to warn that unlimited population growth will magnify global poverty, roll back women's rights, deplete natural resources, and exacerbate the impacts of climate change. Followers can do their part, the group suggests, by supporting family planning policies and limiting the number of children they themselves have.

From 1975 to 1977, the group's national president was John Tanton, an ophthalmologist and a beekeeper in the small northern Michigan town of Petoskey, who also founded local chapters of Planned Parenthood, the Audubon Society, and the Sierra Club. Tanton had tried to push Zero Population Growth to more vocally focus on limiting immigration, but he faced internal pushback.[24] Two years after stepping down as president, Tanton founded the Federation for American Immigration Reform (FAIR) and began formally laying the groundwork for the United States' modern anti-immigration movement.

It is hard to exaggerate the extent to which Tanton reshaped the U.S.—and, by extension, global—political narrative around immigration. He created a system of organizations to fight tooth and nail to reduce the number of foreign-born people allowed into the United States. He was a keen believer in eugenics and race science but effectively cloaked his extremism in a quiet civility. More than any other individual save perhaps Donald Trump, Tanton is responsible for making it publicly acceptable to think that even legal immigration was undermining the country. This was by design. "We plan to make the restriction of immigration a legitimate position for thinking people," he wrote in the group's mission statement, "and to have FAIR identified in the minds of leaders in the media, academia and government as speaking for a consensus of American thought and opinion."

Tanton's development from conservationist to immigration restrictionist occurred several years before the founding of FAIR, illustrated in a 1975 essay submitted as part of the Club of Rome–cosponsored Limits to Growth conference in Woodlands, Texas.[25] The essay, "International Migration as an Obstacle to Achieving World Stability," calls for limiting immigration as a new Malthusian preventive check on over-consumption of resources. "Immigration helps to perpetuate the population and economic growth of the developed nations, which, in turn, will tend to increase their draw on the world's resources," he wrote. "These changes will decrease the amount of food available for export. These are deleterious changes for both the developed and the underdeveloped nations."[26] He went further, claiming that limiting immigration is necessary to force low-income countries to "stabilize."

> As certain portions of the globe deal with their problems more effectively than others, they will stabilize more quickly. This will doubtless increase their attractiveness, especially if other regions are not making progress or are even slipping backwards. This will increase pressures for international migration which, if it is allowed, will tend to destabilize those regions otherwise approaching stability. Thus international migration will have to be stringently controlled, or no region will be able to stabilize ahead of another. If no region can stabilize ahead of another, then it is likely that no region whatsoever will be able to stabilize in an orderly and humane fashion. A more hopeful scenario calls for some regions stabilizing at an early date, and then helping others to do so.
>
> Given the demographic and development situation of the world, the control of international migration will be one of the chief problems the developed countries will face in approaching equilibrium conditions.

FAIR was originally presented as a centrist organization that would steer clear of the bigotry commonly associated with anti-immigration groups. Among its early supporters was the investor Warren Buffett. But as it succeeded in taking immigration restrictionism mainstream, it moved away from the softer touch and began serving a milder version of Tanton's extremism. In private, Tanton insisted that "for European-American society and culture to persist requires a European-American majority, and a clear one at that," and suggested that "minorities" are unable to "run an advanced society." [27] He wrote a paper called "The Case for Passive Eugenics." [28] He rubbed elbows with Holocaust deniers, KKK supporters, and other racists. "As whites see their power and control over their lives declining, will they simply go quietly into the night? Or will there be an explosion?" he once wondered in a memo to like-minded thinkers. [29] FAIR, which the Southern Poverty Law Center has designated a hate group, received more than $1 million from the white supremacist Pioneer Fund and has used its media programming to normalize several white nationalists. [30]

Tanton's project also became an empire. In 1985, FAIR spun off the Center for Immigration Studies (CIS), a think tank designed to make the anti-immigration position seem more scholarly and research-based. The Immigration Reform Law Institute was founded to draft anti-immigration legislation and go to court to support its views. Soon after, Tanton formed a group called U.S. English to fight the government's use of multiple languages on official paperwork and in schools. In the 1990s, Tanton helped shepherd the creation of NumbersUSA to funnel simmering anti-immigrant sentiment into political action to stop congressional immigration reform that might have offered legal status to immigrants who lacked it; the group now describes itself as "the largest online, single-issue, grassroots, advocacy group in America."[31]

Tanton's reach was extensive. From 1966 until his death in 2019, Tanton and his organizations established or substantially funded at least thirty-four different groups focused on immigration reduction, population control, and/or environmental causes.[32] The publishing company he ran republished an English version of the dystopian 1973 French novel *The Camp of the Saints*, in which hordes of immigrants destroy the Western world; the book is now a touchstone of white nationalist circles.[33] As recently as 2004, he sought to resolidify the link between environmentalism and the anti-immigration cause, launching a failed hostile takeover of the Sierra Club's board to force it to embrace immigration restrictions as part of its mandate.[34] Much of the funding for his efforts came from Cordelia Scaife May, an heiress to the Mellon banking fortune whose early political worldview similarly focused on population control and environmental concerns but quickly devolved into nativist fears of an "immigrant invasion."[35] On the surface, this vast network may appear to be entirely unrelated, allowing a passive bystander to mistakenly believe that the U.S. anti-immigration movement grew up entirely organically. In fact, they are a set of interlocking pieces born from the same machine that divides up the work of laundering their message to the public.

Their impact has been profound. In 2007, a young Duke University graduate named Stephen Miller entered Washington as a bullheaded provocateur with an apocalyptic view of U.S. immigration and quickly rose through the ranks on Capitol Hill. He went from the press shop of the firebrand but marginal Republican from Minnesota, Representative Michele

Bachmann, to the office of Senator Jeff Sessions and, eventually, the White House, where he has helped oversee the United States' most anti-immigrant policies in generations. As a top aide to Donald Trump, Miller has had a pivotal behind-the-scenes role in accelerating the United States' embrace of policies to close off access to asylum, ban immigrants from several mostly Muslim-majority countries, and round up and humiliate unauthorized immigrants in the country. He was nurtured along the way by his close ties with groups like FAIR and CIS, whose leaders he knew on a first-name basis and who supplied data and think-tank imprimatur to his once fringe and cruel theories that Black and brown immigrants are undermining the United States and leeching off its bounty.[36]

Tanton himself died in 2019 at the age of eighty-five, after seeing his work bear fruit in Trump's presidency and a historic suite of anti-immigrant policies. Many of them were designed by Miller and others, in consultation with officials who had spent time in the organizations Tanton helped create. Other ideas, such as making English the United States' official language, would come to pass later, when Trump returned to office in 2025. The seeds of Tanton's ideas would continue to sprout for years.

Just a few weeks after he died, a more extreme version of his ideas was wielded by Patrick Crusius, a twenty-one-year-old white nationalist terrorist who unloaded an assault rifle at a Walmart in El Paso, killing twenty-three people in one of the largest single assaults on Latinos in U.S. history. Crusius titled his manifesto "The Inconvenient Truth," mimicking Al Gore's Oscar-winning 2006 climate change documentary, and filled it with rambling claims about a "Hispanic invasion," praise for Dr. Seuss's *The Lorax*, and warnings that overharvesting

U.S. resources was causing environmental devastation. The mass murder of ethnic and racial minorities was necessary to protect the environment, he claimed: "The American lifestyle affords our citizens an incredible quality of life. However, our lifestyle is destroying the environment of our country. The decimation of the environment is creating a massive burden for future generations. . . . I just want to say that I love the people of this country, but god damn most of y'all are just too stubborn to change your lifestyle. So the next logical step is to decrease the number of people in America using resources. If we can get rid of enough people, then our way of life can become more sustainable."[37] Crusius was giving voice to a rising sentiment of eco-fascist hate on the extreme right. In white nationalist circles, climate change has increasingly reared its head as a cause for concern—because of the immigration it drives. "I'm surprised they're not more of these guys," Jared Taylor, a leading proponent of the racist Great Replacement theory that posits that immigrants are subverting white Western culture, later told ProPublica.[38] While Taylor said he disagreed with Crusius's murders, he nonetheless could explain them. Crusius, he said, was "maintaining what is and what is beautiful for the benefit of future generations." There are hundreds of comments about environmentalism and overpopulation on the Nazi-sympathizer website Stormfront, of which 70 percent seem to acknowledge that climate change is real, according to a 2022 study.[39]

The connections are far from solely a U.S. concern. As far-right figures have expanded their voting shares across Europe, they have occasionally acknowledged the existence of climate change and used the issue as a wedge to increase their support. People who are "nomadic," said Marine Le Pen, France's leading far-right politician, "do not care about the environment; they have no homeland." [40] "Borders are the environment's greatest ally," agreed her protégé, Jordan Bardella; "it is through them that we will save the planet." [41]

The reality, it should be made clear, is precisely the opposite of what these voices claim. Generally speaking, a wide array of research has shown that immigrants do not commit crimes at a higher rate than natives and the presence of refugees does not lead to more conflict. [42] Immigrants, particularly those without legal status, tend to be more anxious about running afoul of law enforcement and therefore are disproportionately law-abiding. Sometimes, focus on climate migration can suggest that poor, typically Black and brown people are somehow terrible environmental stewards who mismanage their natural resources and want to steal those of the West as recompense. Alternatively, repeatedly pointing out that poorer countries are historically among the lowest emitters of greenhouse gases treats them as passive characters who have little impact on the world they inhabit. Both lines of thinking are deeply intertwined with condescending ideas about what people in the developing world can and cannot do. One of the arguments of this book is that migration tends not to be a danger, either to the receiving country or to the individual migrant. Although the world has witnessed more migration than ever—only a small portion of it consisting of individuals traveling illicitly—our public debates continue to be enthralled by the idea that immigration is bad and should be stopped.

If Western leaders actually want to reduce immigration,

one option would be to combat climate change and make it easier for people to stay in place. Instead, they have almost uniformly built walls, installed biometric scanners, bought guns, and made immigration ever more difficult—at least for poor people. From 2013 to 2018, seven countries that collectively are responsible for about half of all greenhouse gas emissions (the United States, Canada, the UK, Germany, France, Australia, and Japan) spent at least twice as much money on border security and immigration enforcement as on climate finance.[43] And every year, global spending on border security is higher than the year before. In an average year, the U.S. government spends more or less the equivalent of the entire economy of Iceland ($25 billion) just on immigration enforcement.[44] Meanwhile, borders are being expanded even further. In many respects, the southern border of the United States now begins at the border between Mexico and Guatemala, and the border to the European Union begins in Libya and Morocco, or maybe even farther south. As migration has for a variety of reasons become easier over the last decade and countries contend with historic numbers of irregular migrants, they have frequently relied on so-called transit countries next door to do the work for them, often by dangling development funding, geopolitical support, or some other carrot in front of them. The best place to stop irregular migration, in this view, is well before it reaches the border.

Research suggests that walls mostly do not deter or prevent irregular migration in the long run, although they may redirect it. There is a balloon-like nature to migration, in that squeezing one end focuses the pressure elsewhere. People will generally continue to travel, but they may do so through more remote and therefore more dangerous crossings.[45] That may work just fine for politicians; building a wall is a relatively easy-to-achieve campaign promise in a way that actually halting irregular migration is not. But anyone looking for results that last beyond an election cycle or two will be left wanting.

9

Vacations for the End Times

North Carolina's Highway 12 may be the most climate-threatened road in the United States. It sits awkwardly in the Atlantic several miles off the state's eastern coast, mirroring the coastline like a shadow running a half step behind the rest of the state. It is just two lanes wide for most of its 148 miles, with asphalt surrounded unceremoniously by sand as it winds down a line of touristy barrier islands known as the Outer Banks. As you drive down the highway, low-lying scrub gives way to tall grass, which gives way to pale sand dunes, which give way to ocean. The edges of the road are often covered with wide, shallow puddles of water where the high tide has risen a bit too far to recede back to the ocean.

Tourists spend about $2 billion in the Outer Banks each year and are responsible for about one-third of all jobs.[1] The Outer Banks normally has a population of around 40,000 people, but that swells nearly tenfold in the peak summer months. For long stretches of the Outer Banks, NC-12 is the only road available. On summer weekends, traffic jams stretch for hours as families in bumper-to-bumper congestion begin to second-guess their holiday choices. The road has an official Facebook page, which the state created to provide updates on conditions several times per week; it had about 119,000 followers at the time of this writing.

Eventually, the traffic will clear. It always does. But

tourism—and the highway that makes it possible—is coming under threat because of climate change. Storms have been pounding NC-12 and the Outer Banks with increasing ferocity, turning those isolated roadside puddles into expansive pools. For several years, authorities have erected a wall of sandbags along the shoulder as a stopgap measure to prevent the sea from washing over the highway. In 2022, the state opened a new $155 million bridge to replace a two-mile section of highway near the town of Rodanthe. Instead of continuing overland, along a section of S curves that had been famous in the Outer Banks for disappearing under storm surge during bad weather, the road now swings in over the Pamlico Sound to the west, rising into the air on foundations of solid concrete and positioning the thin strip of barrier island as a barricade. This odd shape has earned it the nickname the "jug handle" bridge, jutting out obtrusively from the otherwise clean, straight line of the island chain. It was the third bridge to be built on NC-12 in five years, as the Outer Banks has raced to protect itself against the sea.

Low-lying barrier islands are by nature a tricky place to build. They shift and evolve with the tides and wind, creating and disappearing small islets as the sand creeps in different directions. On some islands, beaches are eroding by as much as fourteen feet per year.[2] The sea along the Outer Banks is also rising faster than in other parts of the globe, and could rise by more than a foot by 2040 and as much as seven feet by the end of the century, according to National Oceanic and Atmospheric Administration projections.[3] That could wipe away virtually the entire island chain, leaving just a few bare spots of dirt peeking above the waves. The challenges are getting worse. Just a few miles offshore, warm Gulf Stream waters meet the cold Labrador Current, causing chaos and some of the largest waves on the U.S. eastern seaboard. North Carolina's geography resembles a catcher's mitt sticking out into the Atlantic and is particularly prone to being hit by a line drive of storms coming up the eastern seaboard from the Caribbean; as the stitches along the mitt's outer webbing, the barrier islands often bear the brunt. There's a reason why the local high school calls its sports teams the Hurricanes.[4]

Tourists are essential to the Outer Banks' economy, but climate change is slowly pushing them away. It's a conundrum that is afflicting towns and cities across the world's vast archipelago of beachside tourist havens, vacation hot spots, ski slopes, and retirement meccas. As the world heats up, patterns of tourism are changing. Some destinations are becoming less hospitable and some trips are becoming more expensive. Many travelers are avoiding places where the sea is slowly eating away at the highway, where increasingly scorching summers make it impossible to go outside, and where wildfires, hurricanes, and other disasters are becoming the norm. In Europe, three-quarters of travelers have begun adjusting their plans because of climate change.[5] Tourists are also rethinking when they travel, shifting the high season by weeks later or earlier in the year, when the weather is better.

This changing travel is a type of climate-affected movement, too. It may not be as dramatic as the panicked flight of a refugee or a desperate farmer's illicit jump across the border, but tourists are people on the move all the same, and climate change is redirecting whether, where, when, and how they go. For places like the Outer Banks that depend on tourism to prop up the

economy, the shift is potentially existential. Tourism is a massive global industry, responsible for about 10¢ of every dollar
spent worldwide, adding up to trillions of dollars and creating
tens of millions of jobs.[6] Especially after COVID-19, which
created one of the steepest halts to tourism in memory, people
want to travel. But it is a paradox that some of the world's
most desirable geographic locations are also among the most
vulnerable. In fact, it may be precisely because these places
are so vulnerable that they are so desirable, since tourism often depends on natural elements such as pristine beaches, cool
mountain air, or slopes of fresh powder. As tourists reconsider
their travel plans, big companies are starting to rethink how
and where they spend money for the tourism markets of tomorrow, said Daniel Scott, one of the world's leading experts on
sustainable tourism. "The financial community is well aware
of some of these risks and are starting to price that in or build
that into their investment strategies of where are they going to
put that infrastructure, and then, by that definition, where are
we going to try to move tourists to over time?" he told me. "It's
already in motion."[7]

Can the summer vacation as we know it survive?

The Outer Banks is taking steps to hold on. In the town of
Rodanthe, at the southern curve of the jug handle bridge, the
sea is eating away at the houses one by one. In early 2022, just
a few months before the new bridge opened, a house collapsed
along Ocean Drive about a mile south of where the bridge now
extends into the sound.[8] That May, two houses in Rodanthe
collapsed in the span of twenty-four hours.[9] People have tried
to adapt. Both houses were already on stilts towering above the

sand, but these defenses ultimately proved to be futile. When one last wave finally took the houses down, the stilts buckled first, snapped, and collapsed into the surf.[10] They fell clumsily, the bulky homes tottering on spindly legs until they gave out and awkwardly splashed down like a wounded giraffe. Then, slowly, the houses drifted off into the ocean, where they would be dismantled by the waves and scattered up and down the coast for miles. More houses will follow these into the seas. At high tide, the waves already kiss the foot of some houses, begging them to follow their neighbors into the ocean. Seawater seeps into the foundations and trickles onto the street. It's just a matter of time until the houses give in.

To purposefully speed up the destruction, some homes have been bought out by the federal government, in what was likely the first case of the National Park Service buying vulnerable properties just to tear them down. Doing so makes the nearby Cape Hatteras National Seashore "a safer beach," superintendent David Hallac explained, by eliminating teetering houses that posed a risk to vacationers as well as local wildlife.[11] Some owners have physically moved their homes, paying hundreds of thousands of dollars to lift the houses and relocate them a couple of feet farther inland. In the 1990s, the federal government did the same thing with the iconic Cape Hatteras Lighthouse, spending nearly $12 million to lift it and move it half a mile inland to prevent it from sliding into the sea.[12] This movement is a type of the managed retreat discussed earlier in this book: an acknowledgment that the ground on which we have built is no longer suitable, and simply making a plan to get out of the way. Bringing in new sand to shore up the beach would cost $175 million over the next thirty years; buying out the dozens of most at-risk houses would cost a fraction of that.[13]

*   *   *

Moving buildings is not an effective solution to the extreme heat that is proving disastrous for tourism elsewhere. Europe, for instance, is simultaneously the most visited region of the world, accounting for just over half of all international arrivals, and the planet's fastest-warming continent, warming at twice the global average.[14] Sweltering summers are changing precisely what those vacation plans look like. While June and July are still the most popular times to travel to Greece, Italy, and Spain, routinely punishing heat has prompted more travelers to consider the "shoulder season" in the late spring and early autumn, or even in winter months. Tour group organizers, meanwhile, are switching up their daily itineraries to keep tourists inside or by the pool during the day, rather than trekking through ancient ruins under the blazing sun.[15] Sometimes it's not a choice; key sites such as the Acropolis have closed during recent summer afternoons, as temperatures climbed to 110°F.[16]

It's not just unpleasant; hotter summers can also be deadly. Several tourists died during Greece's extreme heat wave in mid-2024, including the British healthy-living TV personality Michael Mosley, whose body was found four days after he disappeared. Temperatures were as high as 104°F on the island of Symi, where Mosley was on vacation, when he set off on a walk and never returned.[17] During heat waves, tourists tend to be particularly vulnerable to severe heat. They are often out and about during the hottest times of the day, walking in the sun in a climate they aren't accustomed to and without the appropriate water or other protections. Tourists also might not know where they're going, which means they're wandering around more than usual while exposed to the sun.

Extreme heat tends to correspond to other disasters that have scared off travelers and put some in peril. In August 2023, a fire broke out in Greece's Evros region near the Turkish border, and over two and a half weeks eventually spread to about 360 square miles of land, causing dozens of villages to be evacuated and registering as the European Union's largest wildfire ever recorded.[18] The blaze was one of hundreds to break out in Greece that summer, with dozens erupting every day. Gale-force winds combined with searing heat and dry weather. The government's climate crisis and civil protection minister, Vassilis Kikilias, called that year's incendiary conditions "the worst since meteorological data have been gathered and the fire risk map has been issued in the country." [19]

Within weeks, Greece's crisis flipped from one of not enough water to one of too much. The dark clouds of early September opened up to drench the country in an unprecedented torrential downpour. Spurred by warmer-than-usual temperatures in the Mediterranean, the storm gathered strength and created flash floods that tore through farming villages and towns. Water raced down hillsides and through streets, dragging along cars, parts of buildings, and anything else left vulnerable to the elements. Tourist areas were rendered inaccessible when roads were washed away, forcing travelers to be rescued by sea.[20] The Mediterranean version of a hurricane (called a "Medicane") known as Storm Daniel beat down on Greece for four days, triggering landslides and bridge collapses in what the mayor of the seaside city of Volos described as a "biblical catastrophe." [21]

When tourists read about events like those, they are more inclined to stay away. One-third of European travelers over age fifty-five say they are avoiding places with extreme temperatures.[22]

*   *   *

Too-warm weather is also a problem for the winter. The Alps now see about one month less snow cover than they used to,[23] and the U.S. snowpack season shrank by an average of eighteen days from 1982 to 2021.[24] The use of snowmaking technologies generally allowed ski resorts to expand their season in recent decades, as they pumped out artificial snow that kept the slopes clean and white. But it takes time for that snow to build up, and many skiers are deciding to wait. "We're definitely seeing a shift in ski seasons," said Daniel Scott, the sustainable tourism researcher. "We're seeing people choose their ski holidays later. So ski areas have more time through the Christmas holidays to develop, with snowmaking, their snowpack, if you go a little bit later. So we're seeing a bit of a shift later into the winter in those cases for sure." And there is a limit to how much snow can feasibly be produced artificially. Making snow is expensive, and when the air temperature sits above freezing for long periods of time, snow just melts away and has to be filled in more frequently.

As a result, seasons are getting not only later but also shorter—and the costs are piling up. "Climate change is the number one threat to the snowsports industry," says the National Ski Areas Association, a major trade group, which has lobbied in support of curbing climate emissions.[25] From 2000 to 2019, climate change cost U.S. ski areas about $5 billion and could cost about $1.4 billion every year by the 2050s, according to one study Scott was involved in.[26] According to that analysis, even in an optimistic scenario, the U.S. ski season will shorten by at least another two weeks by that time; under a more pessimistic, high-emissions scenario, the season will shrink by two months.

Ski resorts might be slightly more able to adapt than other places. Snowmaking technology can help, although only to a point. And anyway, a mountaintop is often beautiful even if it's dry; if resorts can repurpose slopes for other purposes, a new future might be possible. Many ski resorts have started to invest larger amounts of money into hosting warmer-weather activities such as mountain biking, saunas, and ziplining, and also adding opportunities to rent out spaces for weddings and host arts festivals. Canada's Whistler Blackcomb ski resort now sees more visitors in the "green season" than during ski season.[27]

But even then, ski resorts tend to be located in rural, forested areas that climate change is putting at greater risk of wildfires. When the Bridge fire roared through Southern California in autumn 2024, it nearly devastated the Mountain High ski resort, less than two hours from downtown Los Angeles. The resort protected itself in part by strategically deploying snowmaking guns along the periphery, forming a veritable rampart of water to keep the fire out. "When I left out of here," general manager Ben Smith recalled afterward, "I expected to come back to everything gone."[28] Next time, he might be right.

Tourism accounts for about 18 percent of Greece's economy.[29] The Caribbean, where warmer ocean temperatures are bleaching coral reefs and leading to devastating storms that wipe out infrastructure, is more dependent on tourism than any other region of the world, relying on travelers for about one-quarter of the regional economy.[30] In the low-lying Indian Ocean islands of the Maldives, about 40 percent of the economy relies on tourism and related services.[31] What happens to tourist destinations like these if climate change wipes out their major source of revenue?

Australia's "black summer" of 2019 and 2020 might be a preview. That year, a series of bushfires along the country's east ravaged an area as large as the state of Florida, destroying thousands of homes and in some way impacting about four-fifths of all Australians, either directly or indirectly through their friends and family.[32] Black plumes of smoke billowed into the sky and spread across the Southern Hemisphere, lingering in the high altitude for more than a year and a half.[33] The fires helped sink the political career of Prime Minister Scott Morrison, who went on vacation to Hawaii while the blaze raged. And they had a profound effect on Australia's economy, calculated to cost about A$2.8 billion (about $1.8 billion in U.S. dollars) in tourism and related services, adding up to about 7,300 jobs.[34]

Tourism to Australia generally recovered in the years since the massive fires, but that won't be the case everywhere. Smaller countries that are more reliant on tourism might not be able to bounce back so quickly. After twin Hurricanes Irma and Maria plowed through the Caribbean in 2017, about 826,000 fewer tourists decided to come that year, depriving countries of about $741 million in income.[35] To prevent those interruptions from becoming permanent, many countries are eagerly building better infrastructure and trying to adapt, but that too takes money and often plunges countries further into debt. In the three years after severe storms, debt levels in the Caribbean tend to be 18 percent higher than what one would expect otherwise.[36] Meanwhile, the planet is continuing to get warmer and disasters more frequent and more intense.

It's not all doom and gloom, at least not for everybody. In some places, climate change looks like it will be driving more tourism, not less. People still want to go on vacation, and if places like Southern Europe become unattractive, tourists will simply go somewhere else. Summertime travel bookings in places such as Iceland, Scandinavian countries, the Netherlands, and Ireland have risen dramatically in recent years.[37] In 2024, summertime travel to Northern Europe and Canada increased by 44 percent, according to Virtuoso, a luxury travel agency network.[38] A survey by the company found that 82 percent of its clients that year were considering destinations with cooler, moderate weather.[39] Researchers have projected that under a high-emissions scenario, if global temperatures rise by as much as 4°C (7.2°F) by 2100, tourism to parts of Southern Europe could decline by 10 percent but grow in Northern Europe by about 5 percent.[40]

It's unclear whether these "coolcations" are the wave of the future or just a temporary faddish buzzword. Some of the recent increases may be simply a hangover from the COVID-19 pandemic. But countries and private companies have started investing in the idea, and it's clearly a profitable marketing trend. "Norwegian summers are delightfully cool!" a government-sponsored ad campaign boasts. "Lush, green forests. And lots of crisp, fresh air."[41] While Norway's beaches might not be as romantic as those on the Greek islands or the Caribbean, the country's jagged coast means it has one of the world's longest shorelines, with "pristine white sand and turquoise water," according to the promotional campaign. "It looks tropical, although the temperatures might not match that description."

And for a certain type of tourist, the prospect of climate catastrophe is a selling point, not a turnoff. "Last chance tourism" is a niche but growing sector of the global tourism industry in which vacationers try to visit the world's great sites before they vanish forever.[42] Listicles for the "Top Places to Visit Before They Disappear" have become a mainstay of online news sites, particularly travel and outdoors publications. And travelers are listening.

A commonly mentioned spot is icy Churchill, Manitoba, on the western shore of Canada's Hudson Bay. Churchill has a booming tourism industry focused on seeing polar bears in their natural habitat. In the fall, the bears gather near the town as they wait for the ice to freeze over, offering remarkably close-up contact for tourists with a certain verve. Visitors' desire to see the bears has only increased with the sense that the majestic and terrifying beasts may die off, as sea ice disappears in the next few decades. A visit is all the more pressing now, "before they are all gone," as one tourist explained.[43] It's like an inverse bucket list item: trying to visit a place not before you die but before it does. Another top destination is on the exact opposite side of the planet: the ice shelves of Antarctica, where tourism has been increasing steadily for years (not counting the COVID-19 period).[44] And then of course there is Australia's Great Barrier Reef, which is rapidly bleaching to bone white as the seawater gets warmer. More than two-thirds of visitors say a big reason for their trip is "to see the reef before it's gone."[45]

*   *   *

Ironically, no matter where tourists go, they are almost certainly contributing to climate change, thereby accelerating the process they may be trying to outrun. Tourism is a significant contributor to global warming, responsible for as much as 8 percent of global greenhouse gas emissions—more than India or all twenty-seven countries of the European Union.[46] The vast majority of those emissions come from transportation and construction. The climate costs of all those flights, highways, and welcome centers add up. And as the world becomes wealthier and more people are able to take more trips, the amount of emissions they create is rising.[47] So a flight to Scandinavia is in some ways a double blow to warmer hot spots, both depriving them of tourism dollars and increasing the emissions contributing to their undesirability. And the sightseers desperate to catch a glimpse of Hudson Bay polar bears are themselves accelerating the demise of the creatures' habitat just by visiting. Between 2008 and 2018, the total carbon footprint of tourism to Churchill, Canada, increased by 50 percent.[48] Each individual visitor to Antarctica, who usually arrives on a cruise ship, is effectively melting about 83 tons of snow.[49]

Greener travel is possible, but it's not common. In 2019 the Swedish activist Greta Thunberg helped popularize the notion of "flight shame" (*flygskam*), or being embarrassed to take an airplane rather than a train or another mode of transportation. Thunberg has generally taken trains to travel across Europe and that year memorably crossed the Atlantic by sea, in a 60-foot racing yacht, rather than fly. The trend briefly gained traction in Scandinavia. In 2019, more than one-third of Swedes said they would choose to travel by train instead of air, and bookings shifted accordingly.[50] But then came

COVID-19, which prompted a global halt to all movement, and the idea of flight shame got lost in the mix.

Still, it remains the case that a change of habits could help contribute to a greener world, and tourists are, generally speaking, interested in being more mindful. More than two in five travelers across eleven countries said in 2022 that they were actively seeking out more environmentally friendly transportation options, and similar numbers said they were looking to stay in places with a lower environmental impact. Most said they were willing to sacrifice some degree of convenience to travel more sustainably.[51] This is especially true for people who visit endangered areas. More than three-quarters of visitors to the largest glacier in the French Alps, Mer de Glace, afterward said they planned to learn more about protecting the environment, according to a 2020 survey.[52] "It's a place that I love and at the same time makes me very sad," a tourist told the *New York Times*. "I think there's a real opportunity in going there, because you can understand global warming—and feel it."[53]

Tourists in this way sit in a rarified position. They respond to climate change by altering behavior to stay away from or in some cases head toward places on the front lines. They also contribute to climate change every time they step onto a jet. And in seeing the manifestations of climate change up close—and then going home—they also get a firsthand look at its ravages.

10

A New Human Map

The world is on a collision course. Virtually every high-income country is lurching to enact new, dangerous restrictions on immigration by denying access to asylum, shuttering refugee resettlement, and deporting or marginalizing immigrants without legal status. At the same time, the globe is blowing past its emission targets. The United States, historically the world's leading greenhouse gas emitter, has actively gutted its environmental protections, all but ensuring that climate change continues to worsen. These two trends are in direct opposition. Climate change is already destroying neighborhoods, ravaging economies, and forcing people to leave their homes; as the world gets warmer it will continue to do so. Very few people displaced by disasters or moving in response to slow-onset changes will travel internationally, but some of them will. As former Maldives President Mohamed Nasheed outlined in 2014, the world has two choices: "You can drastically reduce your greenhouse gas emissions so that the seas do not rise so much. Or when we show up on your shores in our boats, you can let us in. Or when we show up on your shores in our boats, you can shoot us. You pick." [1]

The age of climate migration has already begun. It will shape this century as profoundly as other types of migration in the early twentieth century shaped that era, restructuring not just our demographics and economics but also the broader human geography of where people live and how. Within the United States, climate change may drive a complex transformation that is for now impossible to predict. Some advocates have marketed the Great Lakes and Rust Belt regions as the United States' climate havens of tomorrow. New boomtowns are emerging in cities like "climate-proof" Duluth, Minnesota, where about 100,000 people live perched at the fingertip of Lake Superior. Buffalo, New York, has launched a promotional campaign around its status as a "climate change refuge."[2] "Buffalo is stepping up and preparing to welcome this new type of refugee," the mayor at the time, Byron Brown, said in his 2019 state of the city address. "We believe that we can accommodate people who have experienced displacement due to harsh weather and natural disaster."[3] After years of decline, these cities are finally starting to grow, due in part to people burned out by blazing summers and endless storms in the South and West. As always, the movement is not without complications. The extra attention is raising home prices in places like Buffalo, which Zillow called the hottest major housing market in both 2024 and 2025.[4] That's good news for current homeowners but bad news for anyone priced out of what was recently a down-and-out city bearing scars of manufacturing's late twentieth-century decline.

We need to think about migration in a new way. There will always be turbulence when new people move in and start changing the economics and character of their new community. But that means we should try to create softer landings, not try to prevent people from coming. Ceasing to move would be against our nature; history shows that migration is the most human of acts. No matter where in the world you are reading this, the reason you are there is because someone at some point migrated. It could have been you, or your grandparents, or at the dawn of humanity, but the fact that it happened is a certainty. Generally speaking, migration is one of the greatest contributors to the world economy, bringing financial benefits to people on both ends of the journey and enriching all of us with a diversity of ideas and stories that create innovative technology and mesmerizing art. It is no coincidence that immigrants play a disproportionately large role in U.S. innovation, or that nearly half of Fortune 500 companies were founded by immigrants or their children.[5] Migration built our world and it will continue to do so in the age of climate change.

But being forced to migrate—whether by the barrel of a gun, or by a collapsing economy, or to outrun a dictator, or by a flood—is traumatic and sometimes deadly. The problem is growing worse, and we are failing to respond. The return of Donald Trump and the rise of his nationalist kin in Europe—France's Marine Le Pen, Germany's Alice Weidel, Italy's Giorgia Meloni, the Netherlands' Geert Wilders, Hungary's Viktor Orbán—are all proof of this. The number of people migrating in distress has outpaced the legal regimes designed to govern their movement and led to a resurgence of nativist leaders who see some of the world's most vulnerable people as a threat. That was true in 2015, when the European Union confronted a record 1.3 million asylum seekers, and it was just as true in 2023, when U.S. authorities encountered nearly 2.5 million asylum seekers and other migrants at the southern border. One out of every sixty-seven people on earth was forcibly displaced as of 2024, nearly double the rate from a decade previous.[6] Most of these people are in the Global South, far from Europe and North America, but all the same, leaders such as Trump have responded to these crises by shutting people out, building walls (literal, legal, and metaphorical), and shrugging their shoulders at the prospect that people either die or live in misery.

This is morally wrong, but it is also shortsighted. An extensive amount of research shows that policies of cruelty have only a short-term effect in stopping people from coming, and that migration eventually returns to its previous level, although people usually end up taking more dangerous and circuitous routes. The fact is that the seas and mountains that used to separate one people from another can now be traversed. Border security can get tougher and tougher, but if climate change (and other factors pushing people to move) gets worse, people will still try to make the journey anyway. This is not necessarily a call for open borders, but the fact of the matter is that, for some of us, open borders essentially already exist. This book is based in part on reporting and research that has taken me to four continents; thanks to my U.S. citizenship, I never faced any restrictions that could not be resolved by waiting in a long line or sending a few polite emails. That is not the case for most people. My navy-blue U.S. passport grants me access to 182 countries without the need to obtain a visa ahead of time. Bangladesh's forest-green passport allows entry to fewer than forty. If you're from Afghanistan, there are only twenty-five countries that will let you in without a thorough review, and those tend to be similarly conflict-stricken and low-income countries like Burundi, Haiti, and Somalia.[7] Meanwhile, the ultra-wealthy can effectively buy legal residence in more than a hundred countries including the United States, Canada, and several in Europe, allowing people with deep pockets to glide past border guards without a second thought.[8]

New legal protections solely for "climate refugees" are probably not the answer. As discussed elsewhere in this book, there are many good and compelling reasons not to simply slap the label "refugees" onto more people, an important one being that individuals themselves don't want it. Neither do host countries: less than one-third of people in the UK say their country has an obligation to protect climate-displaced people, while 41 percent insist it does not.[9] Moreover, there may be a backlash to hyper-fixating on climate migration in the hopes of protecting people. In the United States, reading a lot about climate migration has been shown to have zero effect on whether someone supports policies to combat climate change, and it may lead to more negative attitudes about immigrants.[10]

All the same, the generations-old international refugee system is clearly failing, and we need a new framework. In addition to climate disasters, more people are also fleeing food insecurity, general government collapse, and a range of other issues that fall outside the scope of current international refugee law. Refugee scholar Alexander Betts has coined the phrase "survival migration" to refer to this type of movement; the journalist John Washington has used the phrase "refugees of late capitalism."[11] Whatever the label, the point is that a combination of factors has made more people more vulnerable than ever, and in response they are seeking safety.

Maybe Maldives ex-President Mohamed Nasheed isn't quite right that the only ways out of this mess are to halt climate change or let people migrate freely. There is a third option: Find a way to let people stay in their homes even as the climate conditions deteriorate. Most people do not want to flee,

whether across town or across an ocean, and if we will not mitigate climate change by reducing greenhouse gas emissions, at
least we can help the worst-hit communities adapt to their new
reality. I saw this firsthand in places like Bangladesh, where
fortified seawalls and other defenses in rural villages have allowed people to stay in place and dream of a brighter future
in their current homes. "We basically have three choices: mitigation, adaptation, and suffering," the scientist John Holdren
explained way back in 2007. "We're going to do some of each.
The question is what the mix is going to be. The more mitigation we do, the less adaptation will be required and the less
suffering there will be." [12] Climate change affects everyone, although it does so unevenly, and the countries and communities
most at risk tend to be those that cannot afford to invest in embankments and other adaptations. Seawalls or border walls:
The choice is ours.

We have started in this direction. Remittances are one
way that communities can protect themselves against climate
change, at least in the short term, and the UN's priority to
lower remittance fees could have a significant impact in allowing people to use their own money to protect themselves. In the
longer term, international efforts such as the Adaptation Fund
and the Least Developed Countries Fund have allocated tens
of billions of dollars to help people around the world adapt to
the growing impacts of climate change. [13] The money is still far
short of what is needed—potentially less than one-tenth of total demand, according to the UN Environment Program—but
it is a place to start. [14] "People involved in refugee protection
have a very hard time stopping conflict and civil wars," said
T. Alexander Aleinikoff, a leading refugee law scholar and for-
merly the UN's deputy high commissioner for refugees. "But

in the climate area, you really can begin to talk about helping people stay home and not having to flee, if you build stronger structures to help people withstand storms, or move people to safety a few yards back where the sea-level rise won't occur, or help people with irrigation in drought areas."[15] Aleinikoff has suggested that the international legal regime ought to be reformed to offer people a distinct right not to be displaced, which does not currently exist. Such a protection would allow displaced people to bring legal action against the countries and companies that are disproportionately responsible for climate change, offering the prospect of some sort of justice.

For some places, it is already too late. That is why in 2023 world leaders agreed to create a new Loss and Damage Fund to repay the people and countries where the effects of climate change are irreversible. It is not entirely clear who precisely will benefit from the fund, which is sometimes known as "climate reparations," but in general terms the idea is that high-income, high-polluting countries will provide money to help vulnerable communities recover from losses both economic (such as the need to rebuild destroyed property) and noneconomic (such as the destruction of a culturally important site). For the moment, it remains largely aspirational. The fund had less than $770 million as of mid-2025—a paltry sum, given that climate change regularly contributes to hundreds of billions of dollars in damages each year—and no money has been distributed as of this writing.[16] But if you look for it, there are rays of sunlight peeking through the clouds. A better future is imaginable.

Finally, we also need to reverse the policies that are actively encouraging people to move to dangerous areas. Despite the risks, house prices in places such as the Outer Banks have consistently risen over the past few years, and some of the most climate-vulnerable places in the United States—including in Florida, along the Gulf of Mexico, and on the southeastern Atlantic coast—have actually grown in population.[17] "Some people just want to be on the water no matter what, and/or they want to move here for family, weather, or political reasons," explained a real estate agent in hurricane-prone Cape Coral, Florida. "The Cape is not slowing down." [18]

In the United States, a lot of this building is more or less incentivized by the government. Shortly after Hurricane Betsy ravaged the Gulf states in 1965, the National Flood Insurance Program (NFIP) was created to serve as a type of federal underwriting for building in vulnerable places. It was created with the best of intentions; private companies had largely pulled out of the flood insurance market decades earlier, creating a gap for people in flood zones who were otherwise being ignored. The idea was classic Great Society thinking, in the vein of Medicare or the Head Start program: People collectively can pool their money and, using the federal government as a go-between, distribute it as needed. But the NFIP is also a case study in unintended consequences. Now, virtually every home in flood-prone areas has to carry flood insurance, but the program is deeply in debt because the rates that homeowners pay are not nearly enough to match the payouts given to claimants. In effect, the program has blinded people to climate risk by encouraging them to build in dangerous areas because they know they can get a bailout if things go south. If those risks were priced into the value of a home, instead of being subsidized by the taxpayer, there would be fewer people living in vulnerable deltas, bayous, and barrier islands where they risk being flooded on a regular basis. We are already seeing this transition start to happen with fire risk, as private home insurers cease offering protection or start hiking homeowners' rates. This has created an impossible situation for people already in their homes, who are being pushed out by insurance company diktats, but in years to come it may prevent neighborhoods from building up in dangerous places from which residents will have to flee when wildfires occur.

If there is good news, it is that we are all in this together. Climate change will hit poorer people hardest, but no one is truly spared. The houses in the Outer Banks will sink into the same waves as those in Kiribati. Guatemalan farmers seek the same riches in California as did Dust Bowl Okies a century ago. The scenes presented in this book are not new. Every single human on this planet has been shaped in one way or another by people who migrated in search of riches, safety, family, or just curiosity. When people were primarily hunters and gatherers, we followed animal movements and alluring plants. When we relied mostly on agriculture, we went to wherever had the most fertile soil. When ships and waterways became key to economic growth, we built cities along rivers and bays. And now, as our environment around us is changing faster than ever before, we are confronting a new version of the old tale. When our homes disappear, when our paychecks get smaller, when our neighbors move on without us, and when the prospects of leaving begin to outweigh those of staying in place, we will do what we have always done: We will go somewhere else.

Acknowledgments

First and foremost, I am incredibly grateful to the countless individuals who were willing to speak to me about their lives, experiences, and journeys, sometimes even in traumatic situations. This book is about people, and it would not have been possible without the many kind and selfless individuals who lent me the most precious things of all: their stories.

I am also grateful to the fixers, divers, translators, and other people who helped facilitate my reporting for this book. This includes Riton Quiah, Allen Dias, G. M. Masum Billah, Belal Uddin Joy, Conrado Orellana, Nereyda and Mayra from Asociación de Servicios y Desarrollo Socioeconómico de Chiquimula (ASEDECHI), and Maribel Guzmàn. I also benefitted from the decades of wisdom of scholars, journalists, and development workers, only a small number of whom are quoted or cited in these pages.

Portions of Chapter 7 were financed by a fellowship from the International Reporting Project, which previously served a vital role allowing journalists to tell important stories from around the world. Other portions of that chapter were adapted from my master's thesis at the London School of Economics and Political Science, overseen by Bill Kissane.

The Migration Policy Institute has been my professional home for several years and helped deepen my understanding of how, why, and where people move around the planet. My dedicated colleagues there are thoughtful, caring, and gener-ous, and are helping to enrich the world's migration systems by, among other things, endeavoring to understand them. I would especially like to thank Lawrence Huang for sharing his expertise and Michelle Mittelstadt and Andrew Selee for their professional support.

This is my first book, and even though it is my name on the cover I could not have done it alone. My agent, Malaga Baldi, believed in the idea of this project at a very early stage and was a beacon of light as she guided me through it. Benjamin Woodward had the vision to nurture and mold my text; the en-thusiasm from him and the rest of the team at The New Press bolstered me when I needed it most. Many of my friends pro-vided important emotional and intellectual support through-out this process, and while it would be impossible to list them all I am grateful for each and every one. I am particularly thankful to Claire McNear for offering repeated advice and insight that helped demystify what it means to write a book.

My family has always supported me with love and encour-agement. My parents instilled curiosity in me at a young age, and for all my life my mother has told me to "write the book." Well, here it is. This book is dedicated to Max, who is teaching me more every day about the world as it exists and the world as we would like it to be.

Finally, all of my love and admiration is to Andrea, without whom nothing is worth doing.

Notes

Introduction: Where We Are Going

1. Júlia Ledur, "Visualizing the Scale of the Floods That Left South Brazil Submerged," *Washington Post*, May 23, 2024.

2. Jorge C. Carrasco, "How Climate Displacement Is Affecting Southern Brazil," *El País*, May 14, 2024, https://english.elpais.com/climate/2024-05-14/how-climate-displacement-is-affecting-southern-brazil.html.

3. Jorge C. Carrasco, "Brazil Is Reeling from Catastrophic Floods: What Went Wrong and What Does the Future Hold?" *The Guardian*, May 10, 2024.

4. Lisandra Paraguassu, "Persistent Brazil Floods Raise Specter of Climate Migration," Reuters, May 13, 2024.

5. Internal Displacement Monitoring Centre (IDMC), *2025 Global Report on Internal Displacement* (Geneva: IDMC, 2025), https://www.internal-displacement.org/global-report/grid2025; Shi En Kim, "Six Months After Its Worst Floods, Rio Grande Do Sul Works to Bounce Back," Mongabay, October 28, 2024, https://news.mongabay.com/2024/10/six-months-after-its-worst-floods-rio-grande-do-sul-works-to-bounce-back.

6. "2024 Rio Grande Do Sul, Brazil Floods," Center for Disaster Philanthropy, updated December 4, 2024, https://disasterphilanthropy.org/disasters/2024-rio-grande-do-sul-brazil-floods.

7. Paraguassu, "Persistent Brazil Floods."

8. Carla Ruas, " 'We Have No Tears Left to Cry': Brazilian City Ruined Three Times by Intense Flooding," *A Pública*, May 13, 2024, https://apublica.org/2024/05/we-have-no-tears-left-to-cry-brazilian-city-ruined-three-times-by-intense-flooding.

9. Jorge C. Carrasco, "To Rebuild or Relocate? Southern Brazil's Extreme Floods," *The Dial*, September 17, 2024, https://www.thedial.world/articles/state-of-the-world/southern-brazil-extreme-floods.

10. Kim, "Six Months."

11. Ana Ionova and Tanira Lebedeff, "Images of a Brazilian City Underwater," *New York Times*, May 8, 2024.

12. Marina Dias and Terrence McCoy, "The Climate Refugee Crisis Is Here," *Washington Post*, May 28, 2024.

13. Carrasco, "Brazil Is Reeling."

14. ACAPS, *Bangladesh: Impact of Tropical Cyclone Remal*, briefing note, June 12, 2024, https://www.acaps.org/fileadmin/Data_Product/Main_media/20240612_ACAPS_Bangladesh_-_Impact_of_Tropical_Cyclone_Remal.pdf.

15. Julhas Alam, "A Tropical Storm Floods Villages and Cuts Power to Millions in Parts of Bangladesh and India," Associated Press, May 28, 2024.

16. Jin Yu Young and Saif Hasnat, "Cyclone Remal Tears Through India and Bangladesh, Killing at Least 23," *New York Times*, May 28, 2024.

17. Benoit Nyemba and Ange Adihe Kasongo, "Thousands Homeless after DR Congo's Worst Floods in Sixty Years," Reuters, February 16, 2024.

18. "Record Levels of Flooding in Africa Compounds Stress on Fragile Countries," Africa Center for Strategic Studies, December 10, 2024, https://africacenter.org/spotlight/record-levels-of-flooding-in-africa-compounds-stress-on-fragile-countries.

19. Ruth Maclean and Ismail Alfa, " 'Water Is Coming.' Floods Devastate West and Central Africa," *New York Times*, September 15, 2024.

20. "Storm Boris and European Flooding—September 2024," European Centre for Medium-Range Weather Forecasts, October 23, 2024, https://www.ecmwf.int/en/about/media-centre/focus/2024/storm-boris-and-european-flooding-september-2024.

21. Sophie Tanno, Laura Paddison, Benjamin Brown, and Pau Mosquera, "Spain Hit by Deadliest Floods in Decades. Here's What We Know," CNN, November 1, 2024.

22. Emily Shapiro, David Brennan, Leah Sarnoff, Julia Reinstein, Meredith Deliso, and Ivan Pereira, "Hurricane Helene Updates: Death Toll Surpasses 230 as Rescue Efforts Continue," ABC News, October 7, 2024.

23. National Aeronautics and Space Administration (NASA), "Temperatures Rising: NASA Confirms 2024 Warmest Year on Record," news release, January 10, 2024, https://www.nasa.gov/news-release/temperatures-rising-nasa-confirms-2024-warmest-year-on-record.

24. NASA, "Temperatures Rising."

25. Anna Betts, "Monday Was Hottest Recorded Day on Earth: 'Uncharted Territory,' " *The Guardian*, July 24, 2024.

26. Sarah Kaplan, "Earth Broke All-Time Heat Record Two Days in a Row, Scientists Say," *Washington Post*, July 23, 2024.

27. "Billion-Dollar Weather and Climate Disasters," National Oceanic and Atmospheric Administration (NOAA), National Centers for Environmental Information (NCEI), updated January 10, 2025, https://www.ncei.noaa.gov/access/billions; U.S. Bureau of Economic Analysis (BEA), "Gross Domestic Product by State and Personal Income by State, 3rd Quarter 2024," news release, December 20, 2024, https://www.bea.gov/sites/default/files/2024-12/stgdppi3q24.pdf.

28. IDMC, *2025 Global Report*.

29. UN High Commissioner for Refugees (UNHCR), "Displaced on the Frontlines of the Climate Emergency," ArcGIS StoryMaps, July 19, 2022, https://storymaps.arcgis.com/stories/065d18218b654c798ae9f360a626d903.

30. IDMC, *2025 Global Report*.

31. Bryan McKenzie, "Wet Years Followed by Drought Turned California's Flora into Tinder," University of Virginia Engineering, January 15, 2025, https://engineering.virginia.edu/news-events/news/wet-years-followed-drought-turned-californias-flora-tinder.

32. Yan Zhuang, "The Eaton and Palisades Fires Are Now Among California's Deadliest," *New York Times*, January 27, 2025.

33. Intergovernmental Panel on Climate Change (IPCC), "Summary for Policymakers," in *Climate Change 2022: Impacts, Adaptation and Vulnerability*, ed. H.-O. Pörtner, D.C. Roberts, M. Tignor, E.S. Poloczanska, K. Mintenbeck, A. Alegría, et al. (Cambridge: Cambridge University Press, 2022), https://www.ipcc.ch/report/ar6/wg2/downloads/report/IPCC_AR6_WGII_SummaryForPolicymakers.

34. Rebecca Newman and Ilan Noy, "The Global Costs of Extreme Weather That Are Attributable to Climate Change," *Nature Communications* 14, no. 1 (2023): 6103.

35. "FAQ 2: How Will Nature and the Benefits It Provides to People Be Affected by Higher Levels of Warming?" IPCC, https://www.ipcc.ch/report/ar6/wg2/about/frequently-asked-questions/keyfaq2/.

36. IPCC, "Summary for Policymakers."

37. Viviane Clement, Kanta Kumari Rigaud, Alex de Sherbinin, Bryan Jones, Susana Adamo, Jacob Schewe, et al., *Groundswell Part 2: Acting*

on Internal Climate Migration (Washington, DC: World Bank, 2021), http://hdl.handle.net/10986/36248.

38. "Displacement Data," IDMC, updated May 13, 2025, https://www.internal-displacement.org/database/displacement-data.

39. Lily Katz, "Nearly Half of Americans Who Plan to Move Say Natural Disasters, Extreme Temperatures Factored into Their Decision to Relocate: Survey," Redfin News, April 5, 2021, https://www.redfin.com/news/climate-change-migration-survey.

40. Stewart Macaulay, "Already 3.2 Million Americans Are Climate Migrants," *Governing*, August 22, 2024, https://www.governing.com/climate/already-3-2-million-americans-are-climate-migrants.

41. Lily Katz, Daryl Fairweather, and Sebastian Sandoval-Olascoaga, "Homebuyers with Access to Flood-Risk Data Bid on Lower-Risk Homes," Redfin News, September 12, 2022, https://www.redfin.com/news/redfin-users-interact-with-flood-risk-data.

42. Clement et al., *Groundswell Part 2*.

43. David Wagner, "Asking Rents Skyrocket as LA Fires Destroy Homes," LAist, May 18, 2024, https://laist.com/news/housing-homelessness/los-angeles-palisades-fire-housing-rent-price-gouging-law-california-zillow-listing.

44. "Pacific Palisades Los Angeles, CA Housing Market: 2025 Home Prices & Trends," Zillow, April 30, 2025, https://www.zillow.com/home-values/19810/pacific-palisades-los-angeles-ca.

45. Megan Fan Munce, "L.A. Wildfires Broke Record for Costliest in the History of the Planet," *San Francisco Chronicle*, July 18, 2025.

46. First Street Foundation, *The 12th National Risk Assessment: Property Prices in Peril* (New York: First Street Foundation, February 2025), https://firststreet.org/research-library/property-prices-in-peril.

47. Christopher Flavelle and Mira Rojanasakul, "Insurers Are Deserting Homeowners as Climate Shocks Worsen," *New York Times*, December 18, 2024.

48. Christian Aid, *Human Tide: The Real Migration Crisis* (London: Christian Aid, May 2007), https://reliefweb.int/report/colombia/human-tide-real-migration-crisis-christian-aid-report.

49. Institute for Economics and Peace, "Over One Billion People at Threat of Being Displaced by 2050 Due to Environmental Change, Conflict and Civil Unrest," press release, September 9, 2020, https://www.economicsandpeace.org/wp-content/uploads/2020/09/Ecological-Threat-Register-Press-Release-27.08-FINAL.pdf.

50. Alexander C. Kaufman, "Climate Change Could Threaten Up to 2 Billion Refugees by 2100," *HuffPost*, June 26, 2017, https://www .huffpost.com/entry/climate-change-refugees_n_59506463e4b0da2c731 c5e73.

1. We Have Always Been Climate Migrants

1. Francesco Teo Ficcarelli, Jane Linekar, and Roberto Forin, "Climate-Related Events and Environmental Stressors' Roles in Driving Migration in West and North Africa," Mixed Migration Centre Briefing Paper (Rabat Process, January 2022), https://www.rabat-process.org/en /document-repository/70-studies-publications/389-briefing-paper -climate-change.

2. Robert A. McLeman, *Climate and Human Migration: Past Experiences, Future Challenges* (New York: Cambridge University Press, 2014).

3. Nick A. Drake, Roger M. Blench, Simon J. Armitage, Charlie S. Bristow, and Kevin H. White, "Ancient Watercourses and Biogeography of the Sahara Explain the Peopling of the Desert," *Proceedings of the National Academy of Sciences of the United States of America* 108, no. 2 (2011): 458–462.

4. Amanda Mascarelli, "Climate Swings Drove Early Humans Out of Africa (and Back Again)," *Sapiens*, September 21, 2016, https://www .sapiens.org/biology/early-human-migration.

5. Carl Zimmer, "A Single Migration from Africa Populated the World, Studies Find," *New York Times*, September 21, 2016.

6. Lloyd D. Keigwin, "The Little Ice Age and Medieval Warm Period in the Sargasso Sea," *Science* 274.5292 (1996): 1504–1508.

7. Hubert H. Lamb, "The Early Medieval Warm Epoch and Its Sequel," *Palaeogeography, Palaeoclimatology, Palaeoecology* 1 (1965): 13–37.

8. Brian Fagan, *The Great Warming: Climate Change and the Rise and Fall of Civilizations* (New York: Bloomsbury, 2008).

9. Nicolás E. Young, Avriel D. Schweinsberg, Jason P. Briner, and Joerg M. Schaefer, "Glacier Maxima in Baffin Bay During the Medieval Warm Period Coeval with Norse Settlement," *Science Advances* 1, no. 11 (2015): e1500806.

10. Malcolm K. Hughes and Henry F. Diaz, "Was There a 'Medieval Warm Period,' and if so, Where and When?" *Climatic Change* 26, no. 2 (1994): 109–142; Thomas J. Crowley and Thomas S. Lowery, "How Warm Was the Medieval Warm Period?" *Ambio: A Journal of the Human Environment* 29, no. 1 (2000): 51–54.

11. G. Everett Lasher and Yarrow Axford, "Medieval Warmth Confirmed at the Norse Eastern Settlement in Greenland," *Geology* 47, no. 3 (2019): 267–270.

12. Margot Kuitems, Birgitta L. Wallace, Charles Lindsay, Andrea Scifo, Petra Doeve, Kevin Jenkins, et al., "Evidence for European Presence in the Americas in AD 1021," *Nature* 601, no. 7893 (2022): 388–391.

13. Fagan, *The Great Warming*.

14. Neil Pederson, Amy E. Hessl, Nachin Baatarbileg, and Nicola Di Cosmo, "Pluvials, Droughts, the Mongol Empire, and Modern Mongolia," *Proceedings of the National Academy of Sciences of the United States of America* 111, no. 12 (2014): 4375–4379.

15. Eli Kintisch, "Why Did Greenland's Vikings Disappear?" *Science*, November 10, 2016, https://www.science.org/content/article/why-did-greenland-s-vikings-disappear.

16. Larry V. Benson and Michael S. Berry, "Climate Change and Cultural Response in the Prehistoric American Southwest," *Kiva* 75, no. 1 (2009): 87–117.

17. Barry Pritzker, "Riddles of the Anasazi," *Smithsonian Magazine*, July 2003, https://www.smithsonianmag.com/history/riddles-of-the-anasazi-85274508/; George Johnson, "Vanished: A Pueblo Mystery," *New York Times*, April 8, 2008.

18. Werner Marx, Robin Haunschild, and Lutz Bornmann, "The Role of Climate in the Collapse of the Maya Civilization: A Bibliometric Analysis of the Scientific Discourse," *Climate* 5, no. 4 (2017): 88.

19. G.H. Haug, D. Gunther, L.C. Peterson, D.M. Sigman, K.A. Hughen, and B. Aeschlimann, "Climate and the Collapse of Maya Civilization," *Science* 299, no. 5613 (2003): 1731–1735.

20. Esha Zaveri, Jason Russ, Amjad Khan, Richard Damania, Edoardo Borgomeo, and Anders Jägerskog, *Ebb and Flow*, vol. 1, *Water, Migration, and Development* (Washington, DC: World Bank, 2021), https://hdl.handle.net/10986/36089.

21. David P. Clark, *Germs, Genes, and Civilization: How Epidemics Shaped Who We Are Today* (London: FT Press, 2010).

22. Boris V. Schmid, Ulf Büntgen, W. Ryan Easterday, Christian Ginzler, Lars Walløe, Barbara Bramanti, et al., "Climate-Driven Introduction of the Black Death and Successive Plague Reintroductions into Europe," *Proceedings of the National Academy of Sciences of the United States of America* 112, no. 10 (2015): 3020–3025.

23. Adrian R. Bell, Andrew Prescott, and Helen Lacey, "What Can the Black Death Tell Us About the Global Economic Consequences of a Pandemic?" The Conversation, March 3, 2020, https://theconversation.com/what-can-the-black-death-tell-us-about-the-global-economic-consequences-of-a-pandemic-132793.

24. Christine R. Johnson, "How the Black Death Made Life Better," Washington University in St. Louis, Department of History, June 18, 2021, https://history.wustl.edu/news/how-black-death-made-life-better.

25. Kintisch, "Why Did Greenland's Vikings Disappear?"

26. Linda S. Cordell, Carla R. Van West, Jeffrey S. Dean, and Deborah A. Muenchrath, "Mesa Verde Settlement History and Relocation: Climate Change, Social Networks, and Ancestral Pueblo Migration," *Kiva* 72, no. 4 (2007): 379–405.

27. James Noble Gregory, *American Exodus: The Dust Bowl Migration and Okie Culture in California* (New York: Oxford University Press, 1989).

28. R.A. McLeman, J. Dupre, L. Berrang Ford, J. Ford, K. Gajewski, and G. Marchildon, "What We Learned from the Dust Bowl: Lessons in Science, Policy, and Adaptation," *Population and Environment* 35, no. 4 (2014): 417–440.

29. McLeman et al., "What We Learned from the Dust Bowl."

30. Gregory, *American Exodus*.

31. Richard Hornbeck, "Dust Bowl Migrants: Environmental Refugees and Economic Adaptation," *Journal of Economic History* 83, no. 3 (2023): 645–675.

32. Gregory, *American Exodus*.

33. *The Dust Bowl: A Film by Ken Burns*, directed by Ken Burns (PBS, 2012).

34. Jeffrey A. Lee and Thomas E. Gill, "Multiple Causes of Wind Erosion in the Dust Bowl," *Aeolian Research* 19 (2015): 15–36.

35. Lee and Gill, "Multiple Causes of Wind Erosion."

36. Lee and Gill, "Multiple Causes of Wind Erosion."

37. Matt Shipman, "The Boll Weevil War, or How Farmers and Scientists Saved Cotton in the South," *NC State University News*, May 17, 2017, https://news.ncsu.edu/2017/05/boll-weevil-war-2017.

38. "Cotton Pests," U.S. Department of Agriculture, Animal and Plant Health Inspection Service, updated March 29, 2024, https://www.aphis.usda.gov/plant-pests-diseases/cotton-pests.

39. Gregory, *American Exodus*.

40. Gregory, *American Exodus*.

41. Gregory, *American Exodus*.

42. Gregory, *American Exodus*.

43. Richard A. Warrick, "Drought in the Great Plains: A Case Study of Research on Climate and Society in the USA," *Climatic Constraints and Human Activities* 10 (1980): 93–123.

2. Preparing for a New Atlantis

1. M. Mycoo, D. Campbell, V. Duvat, Y. Golbuu, S. Maharaj, J. Nalau, et al., "Small Islands," in *Climate Change 2022: Impacts, Adaptation and Vulnerability*, ed. H.-O. Pörtner, D.C. Roberts, M. Tignor, E.S. Poloczanska, K. Mintenbeck, A. Alegría, et al. (Cambridge: Cambridge University Press, 2002), https://doi.org/10.1017/9781009325844.017.

2. "Country Rankings—Vulnerability," Notre Dame Global Adaptation Initiative, updated August 26, 2024, https://gain.nd.edu/our-work/country-index/rankings.

3. Simon Albert, Javier X. Leon, Alistair R. Grinham, John A. Church, Badin R. Gibbes, and Colin D. Woodroffe, "Interactions Between Sea-Level Rise and Wave Exposure on Reef Island Dynamics in the Solomon Islands," *Environmental Research Letters* 11, no. 5 (2016): 054011, https://doi.org/10.1088/1748-9326/11/5/054011.

4. Angela Dewan, "Five Solomon Islands Swallowed by the Sea," CNN, May 10, 2016.

5. Denise Chow, "Three Islands Disappeared in the Past Year. Is Climate Change to Blame?" NBC News, June 15, 2019.

6. Robert Oakes, Andrea Milan, and Jillian Campbell, *Kiribati: Climate Change and Migration—Relationships Between Household Vulnerability, Human Mobility and Climate Change* (Bonn: United Nations University Institute for Environment and Human Security, 2016), https://collections.unu.edu/eserv/UNU:5903/Online_No_20_Kiribati_Report_161207.pdf.

7. Médecins Sans Frontières, "Planetary and Public Health Collide in Kiribati," press release, January 19, 2023, https://www.msf.org/kiribati-where-planetary-and-public-health-collide.

8. Alex Kirby, "Islands Disappear Under Rising Seas," BBC News, June 14, 1999.

9. Carly Learson, "Betio Is Facing a Population Crisis, and a Sea Wall Could Be Its Only Hope of Survival," Australian Broadcasting Corpo-

ration, April 5, 2020, https://www.abc.net.au/news/2020-04-05/betio
-facing-overpopulation-crisis-sea-wall-hope-of-survival/11975240.

10. Susin Park, *Climate Change and the Risk of Statelessness: The Situation of Low-Lying Island States* (Geneva: UNHCR, 2011), https://www.unhcr.org/sites/default/files/legacy-pdf/4df9cb0c9.pdf; UNHCR, *Forced Displacement in the Context of Climate Change: Challenges for States Under International Law* (Geneva: UNHCR, 2009), https://www.unhcr.org/us/sites/en-us/files/legacy-pdf/4a1e50082.pdf.

11. Kelly Buchanan, "Tuvalu: Constitutional Amendment Enshrines Statehood in Perpetuity in Response to Climate Change," Library of Congress, September 29, 2023, https://www.loc.gov/item/global-legal -monitor/2023-09-28/tuvalu-constitutional-amendment-enshrines -statehood-in-perpetuity-in-response-to-climate-change.

12. Mycoo et al., "Small Islands."

13. Lena Reimann, Athanasios T. Vafeidis, and Lars E. Honsel, "Population Development as a Driver of Coastal Risk: Current Trends and Future Pathways," *Cambridge Prisms: Coastal Futures* 1 (2023): e14.

14. James Hansen, Makiko Sato, Paul Hearty, Reto Ruedy, Maxwell Kelley, Valerie Masson-Delmotte, et al., "Ice Melt, Sea Level Rise and Superstorms: Evidence from Paleoclimate Data, Climate Modeling, and Modern Observations That 2°C Global Warming Could Be Dangerous," *Atmospheric Chemistry and Physics* 16, no. 6 (2016): 3761–3812.

15. Matthew L. Druckenmiller, Richard L. Thoman, and Twila A. Moon, eds., *Arctic Report Card: Update for 2022* (Silver Spring, MD: NOAA, 2022), https://arctic.noaa.gov/report-card/report-card-2022.

16. Brian Fagan, *The Attacking Ocean: The Past, Present, and Future of Rising Sea Levels* (New York: Bloomsbury, 2014).

17. John Englander, *High Tide on Main Street: Rising Sea Level and the Coming Coastal Crisis* (Boca Raton, FL: Science Bookshelf, 2013).

18. Jeff Goodell, *The Water Will Come: Rising Seas, Sinking Cities and the Remaking of the Civilized World* (New York: Little, Brown, 2017).

19. NASA Science Editorial Team, "Evidence of Sea Level Fingerprints," NASA, September 7, 2017, https://climate.nasa.gov/news/2626/evidence -of-sea-level-fingerprints.

20. United States Geological Survey (USGS), "The Impact of Sea-Level Rise and Climate Change on Pacific Ocean Atolls," USGS Pacific Coastal and Marine Science Center, 2022, https://www.usgs.gov/centers/pcmsc /science/impact-sea-level-rise-and-climate-change-pacific-ocean-atolls.

21. Mathew E. Hauer, Elizabeth Fussell, Valerie Mueller, Maxine Bur-kett, Maia Call, Kali Abel, et al., "Sea-Level Rise and Human Migration," *Nature Reviews Earth & Environment* 1, no. 1 (2020): 28–39.

22. "A Landmark Moment: Tuvalu Is Lifted Above Sea-Level," United Nations Development Programme Pacific Office, November 30, 2023, https://www.undp.org/pacific/stories/landmark-moment-Tuvalu-lifted -above-sea-level.

23. Andrew S. Lewis, "After a Decade of Planning, New York City Is Raising Its Shoreline," Yale Environment 360, December 19, 2023, https:// e360.yale.edu/features/new-york-city-climate-plan-sea-level-rise.

24. Hauer et al., "Sea-Level Rise."

25. Jake Bittle, "Inside the Marshall Islands' Life-or-Death Plan to Survive Climate Change," Grist, December 5, 2023, https://grist.org /extreme-weather/marshall-islands-national-adaptation-plan-sea-level -rise-cop28.

26. Bittle, "Inside the Marshall Islands' Life-or-Death Plan."

27. Julian Hattem, host, *Changing Climate, Changing Migration*, podcast, "Migrate or Adapt? How Pacific Islanders Respond to Cli-mate Change," Migration Policy Institute, February 19, 2021, https:// mpichangingclimatechangingmigration.podbean.com/e/migrate-or-adapt -how-pacific-islanders-respond-to-climate-change.

28. Laurence Caramel, "Besieged by the Rising Tides of Climate Change, Kiribati Buys Land in Fiji," *The Guardian*, June 30, 2014.

29. Ben Doherty, "Climate Change Castaways Consider Move to Aus-tralia," *Sydney Morning Herald*, January 7, 2012.

30. Erica Bower and Sanjula Weerasinghe, *Leaving Place, Restoring Home: Enhancing the Evidence Base on Planned Relocation Cases in the Context of Hazards, Disasters, and Climate Change* (Platform on Disaster Displacement and Andrew & Renate Kaldor Centre for Interna-tional Refugee Law, 2021), https://disasterdisplacement.org/portfolio-item /leaving-place-restoring-home.

31. James R. Elliott and Zheye Wang, "Managed Retreat: A Nationwide Study of the Local, Racially Segmented Resettlement of Homeowners from Rising Flood Risks," *Environmental Research Letters* 18, no. 6 (2023): 064050.

32. John Carter, "Flashback Friday: Niobrara, the Town Too Tough to Stay Put," History Nebraska, Nebraska State Historical Society, 1991, https://history.nebraska.gov/flashback-friday-niobrara-the-town-too -tough-to-stay-put.

33. Nicholas Pinter, "The Lost History of Managed Retreat and Community Relocation in the United States," *Elementa: Science of the Anthropocene* 9 (2021).

34. Carter, "Flashback Friday: Niobrara."

35. Carter, "Flashback Friday: Niobrara."

36. James F. Sterba, "Nebraska Villagers Flee Rising Waters," *New York Times*, May 18, 1974.

37. Carter, "Flashback Friday: Niobrara."

38. Sterba, "Nebraska Villagers Flee Rising Waters."

39. Nebraska State Historical Society, "Wild Weather Wednesday: Niobrara, the Town That Moved Twice," History Nebraska, Nebraska State Historical Society, n.d., https://history.nebraska.gov/wild-weather-wednesday-niobrara-the-town-that-moved-twice.

40. Nicholas Pinter, "True Stories of Managed Retreat from Rising Waters," *Issues in Science and Technology* 37, no. 4 (2021): 64–73.

41. "Isle de Jean Charles Resettlement Project," State of Louisiana, https://isledejeancharles.la.gov.

42. John Connell, "Last Days in the Carteret Islands? Climate Change, Livelihoods and Migration on Coral Atolls," *Asia Pacific Viewpoint* 57, no. 1 (2016): 3–15.

43. Robert A. McLeman, *Climate and Human Migration: Past Experiences, Future Challenges* (New York: Cambridge University Press, 2014).

44. Thibault Le Pivain, "In a 'Noah's Ark' Move, PNG Climate Migrants Bring Thousands of Trees to Safer Ground," Mongabay, November 21, 2024, https://news.mongabay.com/2024/11/in-a-noahs-ark-move-png-climate-migrants-bring-thousands-of-trees-to-safer-ground.

45. Makereta Komai, "Carteret Island Needs K14 Million to Move 350 Families by 2027," Pasifika News, November 24, 2022, https://pasifika.news/2022/11/carteret-island-needs-k14-million-to-move-350-families-by-2027.

46. Katharine J. Mach, Caroline M. Kraan, Miyuki Hino, A. R. Siders, Erica M. Johnston, and Christopher B. Field, "Managed Retreat Through Voluntary Buyouts of Flood-Prone Properties." *Science Advances* 5, no. 10 (2019): eaax8995.

47. Jake Bittle, *The Great Displacement: Climate Change and the Next American Migration* (New York: Simon and Schuster, 2024).

48. Denise M. Bonilla, "Long Island Homeowners Feel Effects of NY's Buyout Program," *Newsday*, October 27, 2016.

49. Kasha Patel, "New York City Is Sinking. These Spots Are Sinking the Fastest," *Washington Post*, September 27, 2023.

50. "Hurricane Hattie 1961," University of the West Indies, https://www.uwi.edu/ekacdm/node/72.

51. "Hurricane 'Hattie' Hits Honduras," British Pathé News Archive, 1961, https://www.britishpathe.com/asset/74857.

52. S. Ricketts, "Belmopan: A New Capital for a New Country," *Docomomo Journal* 43 (2010): 78–81.

53. Juanita Darling, "World Perspective: Government: Belize's Capital Is Clean, Safe—but Few Want to Live There," *Los Angeles Times*, September 11, 1999.

54. Edna Tarigan and Victoria Milko, "Why Is Indonesia Moving Its Capital from Jakarta to Borneo?" Associated Press, March 9, 2023.

55. Hannah Beech, "Welcome to Nusantara: The Audacious Project to Build a Green and Walkable Capital City from the Ground Up," *New York Times*, May 16, 2023.

56. Radio New Zealand, "Marshalls Likens Climate Change Migration to Cultural Genocide," RNZ, October 6, 2015, https://www.rnz.co.nz/news/pacific/286139/marshal's-likens-climate-change-migration-to-cultural-genocide.

57. Sterba, "Nebraska Villagers Flee Rising Waters."

58. "Rwanda Forcefully Evicts Poverty Stricken Kigali Slum Dwellers Without Compensation," *Kampala Post*, December 27, 2018, https://kampalapost.com/content/rwanda-forcefully-evicts-poverty-stricken-kigali-slum-dwellers-without-compensation.

3. After the Flood

1. Yazhou Sun, "Climate Migration Pushes Bangladesh's Megacity to the Brink," Bloomberg News, June 28, 2022.

2. Max Martin, *Climate, Environmental Hazards, and Migration in Bangladesh* (London: Routledge, 2018).

3. Martin, *Climate.*

4. Rashmi Shivni, "Why Is the Indian Ocean Rising So Rapidly?" PBS NewsHour, November 17, 2017, https://www.pbs.org/newshour/science/why-is-the-indian-ocean-rising-so-rapidly.

5. Mohammad Rakibul Hasan, "Bangladesh's Battle Against Climate Change: A Nation at Risk," Inter Press Service, September 1, 2023, https://www.globalissues.org/news/2023/09/01/34638.

6. World Bank, "Population Density (People per Sq. Km of Land Area)," World Bank Group, March 25, 2025, https://data.worldbank.org/indicator/EN.POP.DNST.

7. World Bank, *Poverty & Equity Brief: Bangladesh*, October 2024, https://datacatalogapi.worldbank.org/ddhxext/Resource Download?resource_unique_id=DR0092382.

8. Sohela Mustari, *Climate Induced Migration: Assessing the Evidence from Bangladesh* (Singapore: Partridge, 2022).

9. Refugee and Migratory Movements Research Unit (RMMRU), *Coping with Riverbank Erosion Induced Displacement* (Dhaka: RMMRU, June 2007), http://www.sussex.ac.uk/Units/SCMR/drc/publications/briefing_papers/RMMRU/Policy_brief_ISSUE_1.pdf.

10. Global Center on Adaptation, "Global Center on Adaptation Joins Forces with BRAC to Scale Up Locally Led Adaptation to Build Climate-Resilient, Migrant-Friendly Towns," press release, December 10, 2022, https://gca.org/news/global-center-on-adaptation-joins-forces-with-brac-to-scale-up-locally-led-adaptation-to-build-climate-resilient-migrant-friendly-towns.

11. Sun, "Climate Migration"; "Groundswell: Preparing for Internal Climate Migration," World Bank, March 19, 2018, https://www.worldbank.org/en/news/infographic/2018/03/19/groundswell-preparing-for-internal-climate-migration.

12. Hannah Ritchie, "How Urban Is the World?" Our World in Data, September 27, 2018, https://ourworldindata.org/how-urban-is-the-world.

13. Iman Ghosh, "70 Years of Urban Growth in 1 Dazzling Infographic," World Economic Forum, September 3, 2019, https://www.weforum.org/agenda/2019/09/mapped-the-dramatic-global-rise-of-urbanization-1950-2020.

14. United Nations Department of Economic and Social Affairs, "68% of the World Population Projected to Live in Urban Areas by 2050, Says UN," press release, May 16, 2018, https://www.un.org/development/desa/en/news/population/2018-revision-of-world-urbanization-prospects.html.

15. Raisa Bashar, Sirajus Salekin Tonmoy, Alvira Farheen Ria, and Nazmul Ahsan Khan, "Assessing the Real-Life Socio-Economic Scenario of Established Slums in Dhaka: The Cases of Korail and Sattola," *European Online Journal of Natural and Social Sciences* 9, no. 2 (2020): 455.

16. "Bangladesh—Cyclone Nov 1988 UNDRO Information Reports 1–6," United Nations Department of Humanitarian Affairs, November 30, 1988, https://reliefweb.int/report/bangladesh/bangladesh-cyclone-nov-1988-undro-information-reports-1-6.

17. United Nations Disasters Emergency Committee (DEC), *Bangladesh: 1998 Flood Appeal: An Independent Evaluation* (London: DEC, December 1999), https://reliefweb.int/report/bangladesh/dec-bangladesh -1998-flood-appeal-final-report-independent-evaluation.

18. World Bank, "High Air Pollution Level Is Creating Physical and Mental Health Hazards in Bangladesh, World Bank," press release, December 4, 2022, https://www.worldbank.org/en/news/press-release/2022/12/03 /high-air-pollution-level-is-creating-physical-and-mental-health-hazards -in-bangladesh-world-bank.

19. Thaslima Begum, "Only the Rich Can Bear the Heat: How Dhaka Is Battling Extreme Weather," *The Guardian*, October 3, 2023.

20. Shahadat Hossain, *Urban Poverty in Bangladesh: Slum Communities, Migration, and Social Integration* (New York: I.B. Taurus, 2011).

21. "Bangladesh: Floods Recede, Death Toll Rises," Reuters, September 17, 1998, https://reliefweb.int/report/bangladesh/bangladesh-floods -recede-death-toll-rises.

22. Reaz Haider, "Climate Change-Induced Salinity Affecting Soil Across Coastal Bangladesh," UNB United News of Bangladesh and IPS Inter Press Service, January 15, 2019, https://www.ipsnews.net/2019/01 /climate-change-induced-salinity-affecting-soil-across-coastal-bangladesh.

23. Mustari, *Climate Induced Migration.*

24. Sahana Ghosh, "Climate Change Imperils Sundarbans Tiger Habitats," Mongabay, March 26, 2019, https://india.mongabay.com/2019/03 /climate-change-imperils-sundarbans-tiger-habitats.

25. Asaduzzaman Sardar, "17% of Coastal Population to Become Refugees by 2050," *Dhaka Tribune*, December 27, 2023, https://www .dhakatribune.com/bangladesh/nation/335103/%E2%80%9817%25-of -coastal-population-to-become-refugees-by.

26. International Rescue Committee (IRC), "Bangladesh: IRC Study Reveals Staggering 39% Surge in Child Marriage Due to Climate Change," press release, December 6, 2023, https://www.rescue.org/press-release /bangladesh-irc-study-reveals-staggering-39-surge-child-marriage-due -climate-change.

27 United Nations Population Fund (UNFPA), *Child Marriage and Environmental Crises: An Evidence Brief* (Johannesburg: UNFPA, 2021), https://esaro.unfpa.org/sites/default/files/pub-pdf/child _marriage_and_environmental_crises_an_evidence_brief_final.pdf.

28. Pinaki Roy, "Pratapnagar's Climate Woes Never End," *Daily Star*, November 30, 2023, https://www.thedailystar.net/environment

/climate-crisis/natural-disaster/news/pratapnagars-climate-woes-never
-end-3482196.

29. World Bank, "Employment in Agriculture (% of Total Employment) (Modeled ILO Estimate)," World Bank Group, March 24, 2025, https://data.worldbank.org/indicator/SL.AGR.EMPL.ZS.

30. Philip Gain, *Landless and Social Protection in the Southwest of Bangladesh* (Tala, Bangladesh: Uttaran, 2022).

31. Mustari, *Climate Induced Migration.*

32. Land Governance for Equitable and Sustainable Development (LANDac), *Food Security and Governance Factsheet: Bangladesh* (Utrecht, Netherlands: LANDac, September 2019), https://www.landgovernance.org/wp-content/uploads/2019/09/20160608-Factsheet-Bangladesh.pdf.

33. Barbara Stocking, "Gabura: A Terrifying Vision of a World Devastated by Climate Change," *The Guardian*, November 10, 2009.

34. Abu Azad, "No New Coastal Embankment in 50 Years," *Business Standard*, May 15, 2022, https://www.tbsnews.net/bangladesh/environment/no-new-coastal-embankment-50-years-419954.

35. Tapas Kumar Mollick, *Impact of Climate Change: A Case Study on Gabura Union of Satkhira District, Bangladesh* (Dhaka: Social and Environmental Sustainable Development Organization, March 2023), https://www.researchgate.net/publication/377670581_Impact_of_Climate_Change_A_Case_Study_on_Gabura_Union_of_Satkhira_District_Bangladesh_-_SESDO.

36. Rakibul Hasan, "Bangladesh's Battle."

37. "CEIP-1: Coastal Embankment Improvement Project," Bangladesh Water Development Board, December 2023, http://ceip-bwdb.gov.bd/front.html.

38. "Polder 15: Map Showing Proposed Interventions," Bangladesh Water Development Board, March 21, 2017, http://ceip-bwdb.gov.bd/map/Polder%20Map-15-converted.pdf.

39. Human Rights Forum Bangladesh, *Factsheet—Bangladesh: 4th Cycle Universal Periodic Review, UPR 44—2023—Rights of Indigenous People of Bangladesh* (Dhaka: Human Rights Forum Bangladesh, 2023), https://upr-info.org/sites/default/files/country-document/2023-08/Factsheet-Bangladesh-INDIGENOUS.pdf.

40. Julia Bleckner, "Opinion: Bangladesh's Persecuted Indigenous People," Inter Press Service, May 18, 2015, https://www.ipsnews.net/2015/05/opinion-bangladeshs-persecuted-indigenous-people.

41. Abu Siddique, "In Bangladesh, the Marginalised Munda Face Extra Barriers to Climate Adaptation," *Climate Home News*, November 20, 2020, https://www.climatechangenews.com/2020/11/20/bangladesh -marginalised-munda-face-extra-barriers-climate-adaptation.

4. There Is No Such Thing as a Climate Refugee

1. Ian Goldin, Geoffrey Cameron, and Meera Balarajan, *Exceptional People: How Migration Shaped Our World and Will Define Our Future* (Princeton, NJ: Princeton University Press, 2011).

2. United Nations, *Convention Relating to the Status of Refugees*, July 28, 1951.

3. UNHCR, *Mid-Year Trends 2024* (Copenhagen: UNHCR, 2024), https://www.unhcr.org/mid-year-trends-report-2024.

4. Trinh Tu, "World Refugee Day 2023: Support for the Principle of Refuge Remains High Despite Decline Since 2022," Ipsos, June 19, 2023, https://www.ipsos.com/en-us/world-refugee-day-2023-support-principle -refuge-remains-high-despite-decline-2022.

5. "President Trump Mocks Asylum Seekers, Calls Program a 'Scam,' " C-SPAN, April 6, 2019, https://www.c-span.org/clip/white-house-event /president-trump-mocks-asylum-seekers-calls-program-a-scam/4790668.

6. "Trump's Day-One Actions Attack Legal Pathways to Refuge, Will Harm Core American Values and Interests," Refugees International, January 20, 2017, https://www.refugeesinternational.org/statements-and-news /trumps-day-one-actions-attack-legal-pathways-to-refuge-will-harm-core -american-values-and-interests.

7. UNHCR, *Refugee Statistics*, updated October 8, 2024, https://www .unhcr.org/refugee-statistics/download/?v2url=907f3f.

8. "New Zealand's Net Migration Hits Record High," Australian Associated Press, June 21, 2017, https://www.theguardian.com/world/2017 /jun/22/new-zealand-immigration-hits-record-high.

9. Eleanor Ainge Roy, "New Zealand Labour Signs Coalition Deal and Makes Winston Peters Deputy PM," *The Guardian*, October 24, 2017.

10. Luke Malpass, "Meet Winston Peters, the Donald Trump of the South Pacific," *Australian Financial Review*, September 25, 2017, https:// www.afr.com/world/meet-Winston-peters-the-donald-trump-of-the -south-pacific-20170925-gyo3jj.

11. James Shaw, "Green Party to Welcome 5000 Refugees to New Zealand," Green Party of Aotearoa New Zealand, June 20, 2017, https://www.greens.org.nz/green-party-welcome-5000-refugees-new-Zealand.

12. Radio New Zealand, "NZ Considers Developing Climate Change Refugee Visa," RNZ, October 31, 2017, https://www.rnz.co.nz/international/pacific-news/342749/NZ-considers-developing-climate-change-refugee-visa.

13. "New Zealand Creates Special Refugee Visa for Pacific Islanders Affected by Climate Change," *Straits Times*, December 9, 2017, https://www.straitstimes.com/asia/australianz/new-zealand-creates-special-refugee-visa-for-pacific-islanders-affected-by-climate.

14. Rick Noack, "A Proposal in New Zealand Could Trigger the Era of 'Climate Change Refugees,' " *Washington Post*, October 31, 2017.

15. Ross Giblin, "UNICEF: NZ Must Prepare for an Incoming Tide of Refugees," Stuff, October 9, 2017, https://www.stuff.co.nz/world/97735250/UNICEF-nz-must-prepare-for-an-incoming-tide-of-refugees.

16. Gill Bonnett, "Climate Change Refugee Cases Rejected," RNZ, October 24, 2017, https://www.rnz.co.nz/news/national/342280/climate-change-refugee-cases-rejected.

17. "Tuvalu Climate Change Family Win NZ Residency Appeal," *New Zealand Herald*, August 2, 2014, https://www.nzherald.co.nz/nz/tuvalu-climate-change-family-win-nz-residency-appeal/JMA2SVA2HUB67XTUEXPJXN5DOA.

18. Thomas Manch, "Humanitarian Visa Proposed for Climate Change Refugees 'Dead in the Water,' " Stuff, August 29, 2018, https://www.stuff.co.nz/environment/106660148/humanitarian-visa-proposed-for-climate-change-refugees-dead-in-the-water.

19. Pacific Islands Forum, Leaders' Communiques, Niue Declaration on Climate Change, February 21, 2008, https://forumsec.org/publications/niue-declaration-climate-change.

20. "Pacific Islanders Reject 'Climate Refugee' Status, Want to 'Migrate with Dignity,' SIDS Conference Hears," Australian Broadcasting Corporation, September 5, 2014, https://www.abc.net.au/news/2014-09-05/pacific-islanders-reject-calls-for-27climate-refugee27-status/5723078.

21. Andrew Tillett, "Albanese Opens Borders in Landmark 'Climate Refuge' Deal with Tuvalu," *Australian Financial Review*, November 10, 2023, https://www.afr.com/politics/federal/albanese-opens-borders-in-landmark-climate-refuge-deal-with-tuvalu-20231110-p5ej0i.

22. "PM Speaks on 'Groundbreaking Agreement' Struck between Tuvalu and Australia," *Sky News*, November 10, 2023, https://www.skynews.com.au/world-news/global-affairs/pm-speaks-on-ground breaking-agreement-struck-between-tuvalu-and-australia/video/165eeb d4eb6344770ba1e29262e5122a.

23. Julian Hattem, host, *Changing Climate, Changing Migration*, podcast, "Are the Pacific's Climate Migration Experiments a Preview for the World?" Migration Policy Institute, January 24, 2024, https://mpichangingclimatechangingmigration.podbean.com/e/are-the-pacific -s-climate-migration-experiments-a-preview-for-the-world.

24. "Australia-Tuvalu Falepili Union," Australia Department of Foreign Affairs and Trade (DFAT), September 2024, https://www.dfat.gov.au /sites/default/files/australia-tuvalu-falepili-union-placemat.pdf.

25. Ed Markey, "Senator Markey Introduces First-of-Its-Kind Legislation to Address Climate 'Refugee' Crisis," press release, September 27, 2019, https://www.markey.senate.gov/news/press-releases/senator -markey-introduces-first-of-its-kind-legislation-to-address-climate -refugee-crisis.

26. White House, *Report on the Impact of Climate Change on Migration* (Washington, DC: White House, October 2021), https://bidenwhitehouse .archives.gov/wp-content/uploads/2021/10/Report-on-the-Impact-of -Climate-Change-on-Migration.pdf.

27. United Nations Human Rights Committee, *Report of the Human Rights Committee: 123rd session (2–27 July 2018), 124th session (8 October–2 November 2018), 125th session (4–29 March 2019)* (New York: United Nations, 2019), https://www.un-ilibrary.org/content /books/9789210045858/read.

28. Office of the High Commissioner for Human Rights (OHCHR), "Historic UN Human Rights Case Opens Door to Climate Change Asylum Claims," press release, January 21, 2020, https://www.ohchr.org /en/press-releases/2020/01/historic-UN-human-rights-case-opens-door -climate-change-asylum-claims.

29. Kenneth R. Weiss, "The Making of a Climate Refugee," *Foreign Policy*, January 28, 2015, https://foreignpolicy.com/2015/01/28/the -making-of-a-climate-refugee-kiribati-tarawa-teitiota.

30. Tim McDonald, "The Man Who Would Be the First Climate Change Refugee," BBC News, November 5, 2015.

31. United Nations, *Convention Relating to the Status of Refugees.*

32. Elihu Lauterpacht and Daniel Bethlehem, "The Scope and Content of the Principle of Non-Refoulement," in *Refugee Protection in International*

Law, ed. Erika Feller, Volker Türk, and Frances Nicholson (Cambridge: Cambridge University Press, 2003).

33. United Nations, *Convention Against Torture and Other Cruel, Inhuman or Degrading Treatment or Punishment*, December 10, 1984.

34. Council of Europe, *European Convention for the Protection of Human Rights and Fundamental Freedoms*, November 4, 1950.

35. European Court of Human Rights, *Soering v. United Kingdom*, judgment of July 7, 1989.

36. United Nations General Assembly, *International Covenant on Civil and Political Rights*, December 16, 1966.

37. United Nations Human Rights Committee, *Views Adopted by the Committee Under Article 5 (4) of the of the Optional Protocol, Concerning Communication No. 2728/2016.* September 23, 2020, https://tbinternet.ohchr.org/_layouts/15/treatybodyexternal/Download.aspx?symbolno=CCPR%2fC%2f127%2fD%2f2728%2f2016&Lang=en.

38. Evan Wasuka, "Landmark Decision from UN Human Rights Committee Paves Way for Climate Refugees," Australian Broadcasting Corporation, January 21, 2020, https://www.abc.net.au/news/2020-01-21/un-human-rights-ruling-worlds-first-climate-refugee-kiribati/11887070.

39. Jon Henley, "Man Saved from Deportation After Pollution Plea in French Legal 'First,' " *The Guardian*, January 12, 2021.

40. Corinne Carrière, "Non expulsion d'un 'étranger malade' à Toulouse : Le critère climatique retenu dans la décision de justice," France Info, January 8, 2021, https://france3-regions.francetvinfo.fr/occitanie/haute-garonne/toulouse/non-expulsion-etranger-malade-toulouse-critere-climatique-retenu-decision-justice-1913014.html.

5. Empty Bellies and Empty Pockets

1. Food and Agriculture Organization of the United Nations (FAO), *Land of Opportunities: Dry Corridor in El Salvador, Guatemala and Honduras* (Rome: FAO, 2021), https://www.fao.org/fileadmin/user_upload/rlc/docs/DryCorridor.pdf.

2. John L. Beven III, *Tropical Cyclone Report: Tropical Storm Agatha (EP012010)* (Miami: National Hurricane Center, 2010), https://www.nhc.noaa.gov/data/tcr/EP012010_Agatha.pdf.

3. Rachel Schmidtke and Kayly Ober, *Two Years After Eta and Iota: Displaced and Forgotten in Guatemala* (Washington, DC: Refugees

International, 2023), https://www.refugeesinternational.org/reports-briefs /two-years-after-eta-and-iota-displaced-and-forgotten-in-guatemala.

4. Marilia Brocchetto and Nelson Quinones, "Guatemala Landslide Death Toll Rises to 271," CNN, October 10, 2015.

5. "Hundreds Still Missing After Deadly Landslide in Guatemala; Authorities Doubtful of Finding More Survivors," Australian Broadcasting Corporation, October 3, 2015, https://www.abc.net.au/news/2015-10-04 /guatemala-landslide-hopes-of-finding-survivors-fading/6826252.

6. Terrence McCoy, "Brazil Mudslides: Climate Change Turns Favelas into Disasters Waiting to Happen," *Washington Post*, February 24, 2022.

7. World Food Program (WFP), "Princess Sarah Zeid of Jordan Calls for Women's Empowerment During Visit to WFP Programmes in Guatemala," press release, July 26, 2023, https://www.wfp.org/news /princess-sarah-zeid-jordan-calls-womens-empowerment-during-visit -wfp-programmes-guatemala.

8. World Bank, "Employment in Agriculture (% of Total Employment)— Guatemala," World Bank Group, updated January 7, 2025, https://data .worldbank.org/indicator/SL.AGR.EMPL.ZS?locations=GT.

9. Instituto Nacional de Estadística Guatemala (INE), *XII censo nacional de población y VII de vivienda: Principales resultados censo 2018* (Guatemala City: INE, 2019), https://www.censopoblacion.gt/archivos /resultados_censo2018.pdf.

10. Paris Rivera, Werner Ochoa, and Marvin Salguero, *Escenarios de cambio climático para Guatemala* (Guatemala City: Universidad de San Carlos de Guatemala, 2020), https://sgccc.org.gt/wp-content /uploads/2020/09/ESCENARIOS-DE-CAMBIO-CLIMATICO-PARA -GUATEMALA-Agosto-Final.pdf.

11. WFP, *Food Security and Emigration: Why People Flee and the Impact on Family Members Left Behind in El Salvador, Guatemala and Honduras* (Panama City: WFP Regional Bureau for Latin America and the Caribbean, 2017), https://reliefweb.int/report/el-salvador/food -security-and-emigration-why-people-flee-and-impact-family-members -left.

12. Ariel G. Ruiz Soto, Rossella Bottone, Jaret Waters, Sarah Williams, Ashley Louie, and Yuehan Wang, *Charting a New Regional Course of Action: The Complex Motivations and Costs of Central American Migration* (Washington, DC: Migration Policy Institute, 2021), https:// www.migrationpolicy.org/research/motivations-costs-central-american -migration.

13. Andrew Linke, Stephanie Leutert, Joshua Busby, Maria Duque, Matthew Shawcroft, and Simon Brewer, "Dry Growing Seasons Predicted Central American Migration to the US from 2012 to 2018," *Scientific Reports* 13, no. 1 (2023): 18400.

14. Ana Maria Ibañez, Juliana Quigua, Jimena Romero, and Andrea Velásquez, *Responses to Temperature Shocks: Labor Markets and Migration Decisions in El Salvador* (Washington, DC: Inter-American Development Bank, 2022), https://publications.iadb.org/en/publications /english/viewer/Responses-to-Temperature-Shocks-Labor-Markets-and -Migration-Decisions-in-El-Salvador.pdf.

15. Susanne Jonas, "Guatemalan Migration in Times of Civil War and Post-War Challenges," *Migration Information Source*, March 27, 2013, https://www.migrationpolicy.org/article/Guatemalan-migration-times -civil-war-and-post-war-challenges.

16. Susanne Jonas and Nestor Rodríguez, *Guatemala-U.S. Migration: Transforming Regions* (Austin: University of Texas Press, 2015).

17. Commission for Historical Clarification (CEH), *Guatemala: Memory of Silence—Tz'inil Na 'Tab'Al—Report of the Commission for Historical Clarification: Conclusions and Recommendations* (Guatemala City: CEH, 1999), https://hrdag.org/wp-content/uploads/2013/01/CEHreport -english.pdf.

18. Stephen Schlesinger and Stephen Kinzer, *Bitter Fruit: The Story of the American Coup in Guatemala* (Cambridge, MA: Harvard University Press, 2005).

19. Aviva Chomsky, *Central America's Forgotten History: Evolution, Violence, and the Roots of Migration* (Boston: Beacon Press, 2021).

20. Greg Grandin, *The Blood of Guatemala: A History of Race and Nation* (Durham, NC: Duke University Press, 2000).

21. Jonas and Rodríguez, *Guatemala-U.S. Migration*.

22. Chomsky, *Central America's Forgotten History*.

23. Jonas and Rodríguez, *Guatemala-U.S. Migration*.

24. "Countries of Birth for U.S. Immigrants, 1960–Present," Migration Data Hub, Migration Policy Institute, https://www.migrationpolicy.org /programs/data-hub/us-immigration-trends.

25. "Top Diaspora Groups in the United States, 2023," Migration Data Hub, Migration Policy Institute, https://www.migrationpolicy.org /programs/data-hub/charts/top-diaspora-groups-united-states-2023.

26. Jonas and Rodríguez, *Guatemala-U.S. Migration*.

27. Jonathan Blitzer, *Everyone Who Is Gone Is Here: The United States, Central America, and the Making of a Crisis* (New York: Penguin Books, 2025).

28. Ruiz Soto et al., *Charting a New Regional Course of Action.*

29. Ruiz Soto et al., *Charting a New Regional Course of Action.*

30. Jennifer Van Hook, Ariel G. Ruiz Soto, and Julia Gelatt, "The Unauthorized Immigrant Population Expands amid Record U.S.-Mexico Border Arrivals," Migration Policy Institute, February 2025, https://www.migrationpolicy.org/news/unauthorized-immigrant-population-mid-2023.

31. Transactional Records Access Clearinghouse (TRAC), *The Impact of Nationality, Language, Gender and Age on Asylum Success* (Syracuse, NY: TRAC, 2021), https://tracreports.org/immigration/reports/668.

32. Nicole Ward and Jeanne Batalova, "Central American Immigrants in the United States," *Migration Information Source*, May 10, 2023, https://www.migrationpolicy.org/article/central-american-immigrants-united-states-2021.

33. Victoria A. Greenfield, Blas Nuñez-Neto, Ian Mitch, Joseph C. Chang, and Etienne Rosas, *Human Smuggling and Associated Revenues: What Do or Can We Know About Routes from Central America to the United States?* (Santa Monica, CA: RAND, 2019), https://www.rand.org/pubs/research_reports/RR2852.html.

34. Sonia Pérez D., "Guatemala Follows Money in Migrant Smuggling Investigations," Associated Press, August 23, 2022.

35. World Bank, "Urban Population (% of Total Population)—Guatemala," World Bank Group, updated March 24, 2025, https://data.worldbank.org/indicator/SP.URB.TOTL.IN.ZS?locations=GT.

36. World Bank, *Groundswell: Preparing for Internal Climate Migration—Policy Note #3: Internal Climate Migration in Latin America* (Washington, DC: World Bank, 2018), https://documents1.worldbank.org/curated/en/983921522304806221/pdf/124724-BRI-PUBLIC-NEW SERIES-Groundswell-note-PN3.pdf.

37. Kanta Kumari Rigaud, Alex de Sherbinin, Bryan Jones, Jonas Bergmann, Viviane Clement, Kayly Ober, et al., *Groundswell: Preparing for Internal Climate Migration* (Washington, DC: World Bank, 2018), http://hdl.handle.net/10986/29461.

6. Migration Can Be a Solution— for People Who Can Get Out

1. Joseph Shapiro, "Disaster Relief for the Elderly and Disabled Is Already Hard. Now Add a Pandemic," NPR, July 22, 2020, https://www.npr.org/2020/07/22/894148776/disaster-relief-for-the-elderly-and-disabled-is-already-hard-now-add-a-pandemic.

2. "Missing Migrants and Countries in Crisis," International Organization for Migration (IOM), April 28, 2025, https://storymaps.arcgis.com/stories/1098aa8ecb07417ab4276607092149cc.

3. Isabella Lloyd-Damnjanovic, "Criminalization of Search-and-Rescue Operations in the Mediterranean Has Been Accompanied by Rising Migrant Death Rate," *Migration Information Source*, October 9, 2020, https://www.migrationpolicy.org/article/criminalization-rescue-operations-Mediterranean-rising-deaths.

4. Michel Beine, Ilan Noy, and Christopher Parsons, *Climate Change, Migration and Voice: An Explanation for the Immobility Paradox* (Bonn: Institute of Labor Economics, 2019), https://www.iza.org/publications/dp/12640/climate-change-migration-and-voice-an-explanation-for-the-immobility-paradox.

5. Nomaan Merchant and Sonia Pérez D., "US Won't Answer New Questions About Migrant Teen's Death," Associated Press, May 9, 2019.

6. Jeff Abbott, "I Lost My Son: Guatemala Mum Mourns Boy Who Died in US Custody," Al Jazeera, May 3, 2019; Nina Lakhani, "Guatemalan Boy Who Died in US Custody Suffered from Brain Infection," *The Guardian*, May 3, 2019.

7. Beine et al., *Climate Change, Migration and Voice.*

8. Merewalesi Yee, Karen E. McNamara, Annah E. Piggott-McKellar, and Celia McMichael, "The Role of Vanua in Climate-Related Voluntary Immobility in Fiji," *Frontiers in Climate* 4 (2022): 1034765.

9. Briana Nichols, "Nothing Is Easy: Educational Striving and Migration Deferral in Guatemala," *Journal of Ethnic and Migration Studies* 49, no. 7 (2023): 1919–1935.

10. June Cross, "The Old Man and the Storm," PBS Frontline, January 6, 2009, https://www.pbs.org/wgbh/pages/frontline/katrina/etc/script.html.

11. Çaglar Özden and Mathis Wagner, *Moving for Prosperity: Global Migration and Labor Markets* (Washington, DC: World Bank, 2018), https://www.worldbank.org/en/research/publication/moving-for-prosperity.

12. John Gibson, David McKenzie, Halahingano Rohorua, and Steven Stillman, "The Long-Term Impacts of International Migration: Evidence from a Lottery," *World Bank Economic Review* 32, no. 1 (2018): 127–147.

13. Michael Clemens, *The Emigration Life Cycle: How Development Shapes Emigration from Poor Countries* (Washington, DC: Center for Global Development, 2020), https://www.cgdev.org/publication/emigration -life-cycle-how-development-shapes-emigration-poor-countries.

14. World Bank, "GDP per Capita (Current US$)–Mexico," World Bank Group, updated March 23, 2025, https://data.worldbank.org/indicator /NY.GDP.PCAP.CD?locations=MX; Jeanne Batalova, "Mexican Immigrants in the United States," *Migration Information Source*, October 4, 2024, https://www.migrationpolicy.org/article/mexican-immigrants -united-states-2024.

15. Özden and Wagner, *Moving for Prosperity*.

16. Ruiz Soto et al., *Charting a New Regional Course of Action*.

17. Ruiz Soto et al., *Charting a New Regional Course of Action*.

18. "Guatemala's Exports Superseded by Remittances in 2022," Agence France-Presse, January 7, 2023, https://www.barrons.com/news /Guatemala-s-exports-superceded-by-remittances-in-2022-01673120409.

19. Justice Issah Musah-Surugu and Samuel Weniga Anuga, "Remittances as a Game Changer for Climate Change Adaptation Financing for the Most Vulnerable: Empirical Evidence from Northern Ghana," in *Remittances as Social Practices and Agents of Change: The Future of Transnational Society*, ed. Silke Meyer and Claudius Ströhle, 343–367 (Cham: Springer International Publishing, 2023).

20. Sanket Mohapatra, George Joseph, and Dilip Ratha, "Remittances and Natural Disasters: Ex-Post Response and Contribution to Ex-Ante Preparedness," *Environment, Development and Sustainability* 14, no. 3 (2012): 365–387.

21. Teresa Randazzo, Filippo Pavanello, and Enrica De Cian, "Adaptation to Climate Change: Air-Conditioning and the Role of Remittances," *Journal of Environmental Economics and Management* 120 (2023): 102818.

22. Goldin et al., *Exceptional People*.

23. Dilip Ratha, Sonia Plaza, and Eung Ju Kim, "In 2024, Remittance Flows to Low- and Middle-Income Countries Are Expected to Reach $600 Billion, Larger than FDI and ODA Combined," *World Bank Blogs*, December 18, 2024, https://blogs.worldbank.org/en/peoplemove /in-2024—remittance-flows-to-low—and-middle-income-countries-ar.

24. Gumisai Mutume, "Workers' Remittances: A Boon to Development," *Africa Renewal*, October 2005, https://www.un.org/africarenewal/magazine/october-2005/workers%E2%80%99-remittances-boon-development.?

25. Richard H. Adams Jr. and John Page, "Do International Migration and Remittances Reduce Poverty in Developing Countries?" *World Development* 33, no. 10 (2005): 1645–1669.

26. Ruiz Soto et al., *Charting a New Regional Course of Action.*

27. David Kestenbaum, "What Happens When You Just Give Money to Poor People?" NPR, October 25, 2013, https://www.npr.org/sections/money/2013/10/25/240590433/what-happens-when-you-just-give-money-to-poor-people.

28. Samuel Huckstep and Michael Clemens, *Climate Change and Migration: An Overview for Policymakers and Development Practitioners* (Washington, DC: Center for Global Development, 2023), https://www.cgdev.org/publication/climate-change-and-migration-overview-policymakers-and-development-practitioners.

29. Jeanne Batalova, "Immigrant Health-Care Workers in the United States," *Migration Information Source*, April 7, 2023, https://www.migrationpolicy.org/article/immigrant-health-care-workers-united-states-2021.

30. Maruja M.B. Asis, "The Philippines: Beyond Labor Migration, Toward Development and (Possibly) Return," *Migration Information Source*, July 12, 2017, https://www.migrationpolicy.org/article/philippines-beyond-labor-migration-toward-development-and-possibly-return.

31. Ana P. Santos, "Philippines: 400,000 Seafarers at Risk of Sailing Ban," DW, December 26, 2022, https://www.dw.com/en/philippines-400000-seafarers-at-risk-of-sailing-ban/a-64213556.

32. "Labour Migration in the Philippines," International Labour Organization, https://www.ilo.org/manila/areasofwork/labour-migration/lang—en/index.htm.

33. "Global Remittances Guide," Migration Data Hub, Migration Policy Institute, https://www.migrationpolicy.org/programs/data-hub/global-remittances-guide.

34. Reichert, Joshua S. Reichert, "The Migrant Syndrome: Seasonal US Wage Labor and Rural Development in Central Mexico," *Human Organization* 40, no. 1 (1981): 56–66.

35. Ruiz Soto et al., *Charting a New Regional Course of Action.*

36. World Bank, "Personal Remittances, Received (% of GDP)," World Bank Group, updated March 23, 2025, https://data.worldbank.org/indicator/BX.TRF.PWKR.DT.GD.ZS.

37. Edward Lemon, "Dependent on Remittances, Tajikistan's Long-Term Prospects for Economic Growth and Poverty Reduction Remain Dim," *Migration Information Source*, November 14, 2019, https://www.migrationpolicy.org/article/dependent-remittances-tajikistan-prospects-dim-economic-growth.

38. Zubaidah Abdul Jalil, "Tonga: How an Internet Blackout Left Many Desperate for Money," BBC News, February 6, 2022.

39. Ryan Edwards, Matthew Dornan, Dung Doan, and Toan Nguyen, "Tonga: Three Questions on Tongan Remittances," *Devpolicy Blog*, July 20, 2022, https://devpolicy.org/three-questions-on-tongan-remittances-20220720.

40. World Bank, "Personal Remittances, Received (% of GDP)."

7. Outrunning Bullets and Disasters

1. Hala Gorani, Briony Sowden, and Astha Rajvanshi, "The Syrian Teenager Who Sprayed Four Words on a Wall and Started an Uprising," NBC News, February 23, 2025.

2. Philip Loft and Esme Kirk-Wade, "The Syrian Civil War: Timeline, UK Aid and Statistics," House of Commons Library, December 19, 2024, https://commonslibrary.parliament.uk/research-briefings/cbp-9381.

3. Massoud Ali, *Years of Drought: A Report on the Effects of Drought on the Syrian Peninsula* (Beirut: Heinrich-Böll-Stiftung, 2010), https://lb.boell.org/sites/default/files/uploads/2010/12/drought_in_syria_en.pdf.

4. Colin P. Kelley, Shahrzad Mohtadi, Mark A. Cane, Richard Seager, and Yochanan Kushnir, "Climate Change in the Fertile Crescent and Implications of the Recent Syrian Drought," *Proceedings of the National Academy of Sciences of the United States of America* 112, no. 11 (2015): 3241–3246.

5. Thomas L. Friedman, "Without Water, Revolution," *New York Times*, May 18, 2013.

6. Kelley et al., "Climate Change in the Fertile Crescent."

7. Kelley et al., "Climate Change in the Fertile Crescent."

8. Marwa Daoudy, *The Origins of the Syria Conflict: Climate Change and Human Security* (Cambridge: Cambridge University Press, 2020).

9. Friedman, "Without Water, Revolution."

10. Marshall Burke, Solomon M. Hsiang, and Edward Miguel, "Climate and Conflict," *Annual Review of Economics* 7, no. 1 (2015): 577–617.

11. Kate Burrows and Patrick L. Kinney, "Exploring the Climate Change, Migration and Conflict Nexus," *International Journal of Environmental Research and Public Health* 13, no. 4 (2016): 443.

12. Rafael Reuveny, "Climate Change–Induced Migration and Violent Conflict," *Political Geography* 26, no. 6 (2007): 656–673.

13. *Before the Flood*, directed by Fisher Stevens (Beverly Hills, CA: RatPac Documentary Films, 2016).

14. Peter Schwartzstein, *The Heat and the Fury: On the Frontlines of Climate Violence* (Washington, DC: Island Press, 2024).

15. Christopher Cramer, *Civil War Is Not a Stupid Thing: Accounting for Violence in Developing Countries* (London: Hurst, 2006).

16. Samuel J. Michel, Han Wang, Shalini Selvarajah, Joseph K. Canner, Matthew Murrill, Albert Chi, et al., "Investigating the Relationship Between Weather and Violence in Baltimore, Maryland, USA," *Injury* 47, no. 1 (2016): 272–276; Paul M. Reeping and David Hemenway, "The Association Between Weather and the Number of Daily Shootings in Chicago (2012–2016)," *Injury Epidemiology* 7, no. 1 (2020): 1–8.

17. Anita Mukherjee and Nicholas J. Sanders, "The Causal Effect of Heat on Violence: Social Implications of Unmitigated Heat Among the Incarcerated," working paper no. w28987 (National Bureau of Economic Research, 2021).

18. V.H. Lyons, E.L. Gause, K.R. Spangler, G.A. Wellenius, and J. Jay, "Analysis of Daily Ambient Temperature and Firearm Violence in 100 US Cities," *JAMA Network Open* 5, no. 12 (2022): e2247207.

19. Ben Taub, "Lake Chad: The World's Most Complex Humanitarian Disaster," *New Yorker*, November 27, 2017, https://www.newyorker.com/magazine/2017/12/04/lake-chad-the-worlds-most-complex-humanitarian-disaster.

20. Jennifer Epstein, "Michelle Obama's Hashtag Gamble," Politico, May 13, 2014, https://www.politico.com/story/2014/05/michelle-obama-bringbackourgirls-106645.

21. Institute for Economics and Peace, *Global Terrorism Index 2015: Measuring and Understanding the Impact of Terrorism* (Sydney: Institute for Economics and Peace, 2015), https://www.economicsandpeace.org/wp-content/uploads/2015/11/Global-Terrorism-Index-2015.pdf.

22. "Boko Haram," National Counterterrorism Center, updated October 2022, https://www.dni.gov/nctc/ftos/boko_haram_fto.html.

23. Leif Brottem, *The Growing Complexity of Farmer-Herder Conflict in West and Central Africa* (Washington, DC: Africa Center for Strategic Studies, 2021), https://africacenter.org/publication/growing-complexity -farmer-herder-conflict-west-central-africa.

24. Janani Vivekananda, Martin Wall, Florence Sylvestre, and Chitra Nagarajan, *Shoring Up Stability: Addressing Climate and Fragility Risks in the Lake Chad Region* (Berlin: Adelphi, 2019), https://shoring-up -stability.org/the-report.

25. Vivekananda et al., *Shoring Up Stability*.

26. M.A.P. Montclos, "Boko Haram, Youth Mobilization and Jihadism," in *Creed and Grievance: Muslim-Christian Relations and Conflict Resolution in Northern Nigeria*, ed. Abdul Raufu Mustapha and David Ehrhardt (Suffolk, UK: James Currey, 2018), 165–183.

27. Paul Carsten and Ahmed Kingimi, "Islamic State Ally Stakes Out Territory Around Lake Chad," Reuters, April 29, 2018.

28. Gillian Parker, "Nigeria's Abandoned Youth: Are They Potential Recruits for Militants?" *Time*, February 18, 2012.

29. Wole Soyinka, "The Butchers of Nigeria," *Newsweek*, January 16, 2012, https://www.newsweek.com/wole-soyinka-nigerias-anti-christian -terror-sect-boko-haram-64153.

30. "More than 100 Killed as Herders, Farmers Clash in Central Nigeria," France 24, May 19, 2023, https://www.france24.com/en/africa/20230519 -more-than-100-killed-as-herders-farmers-clash-in-central-nigeria.

31. Chinedu Asadu, "At Least 80 People Were Killed in an Attack in Northern Nigeria. Police Arrested 7 Suspects," Associated Press, May 18, 2023.

32. Schwartzstein, *The Heat and the Fury*.

33. UNHCR, *No Escape: On the Frontlines of Climate Change, Conflict, and Forced Displacement* (Geneva: UNHCR, 2022), https://www .unhcr.org/media/no-escape-frontlines-climate-change-conflict-and -forced-displacement.

34. UNHCR, *No Escape*.

35. "Myanmar: Crimes Against Humanity Committed Systematically, Says UN Report," UN News, August 9, 2022, https://news.un.org/en /story/2022/08/1124302.

36. OHCHR, *Report of OHCHR Mission to Bangladesh: Interviews with Rohingyas Fleeing from Myanmar Since 9 October 2016* (Geneva: OHCHR, 2017), https://www.ohchr.org/sites/default/files/Documents /Countries/MM/FlashReport3Feb2017.pdf.

37. "Deadly Cyclone Mora Hits Bangladesh with High Winds and Rain," BBC News, May 30, 2017.

38. UNHCR, *No Escape.*

39. Kaamil Ahmed, " 'We Are Going to Die Collecting Firewood,' " Roads & Kingdoms, October 18, 2018, https://roadsandkingdoms .com/2018/photo-essay-rohingya-firewood.

40. Julian Hattem, host, *Changing Climate, Changing Migration*, podcast, "No 'Climate Refugees,' but Still a Role for the UN Refugee Agency" Migration Policy Institute, April 19, 2021, https:// mpichangingclimatechangingmigration.podbean.com/e/no-climate -refugees-but-still-a-role-for-the-un-refugee-agency/.

41. UNHCR, *No Escape.*

42. M. Sanjeeb Hossain and Maja Janmyr, "Bhasan Char: Prison Island or Paradise? Are Rohingya Refugees Being Denied Their Right to Freedom of Movement?" *Lacuna*, May 12, 2022, https://lacuna.org.uk/migration /bhasan-char-rohingya-refugees-right-to-freedom-of-movement.

8. Population Panic and the Environmentalist Seeds of the Modern Anti-Immigration Movement

1. Lloyd Austin, "Secretary Austin Remarks at Climate Change Summit," April 22, 2021, U.S. Department of Defense, transcript, https://www .defense.gov/News/Transcripts/Transcript/Article/2582828/secretary -austin-remarks-at-climate-change-summit.

2. Robert D. Kaplan, "The Coming Anarchy," *The Atlantic*, February 1994, https://www.theatlantic.com/magazine/archive/1994/02/the -coming-anarchy/304670.

3. Andrew Revkin, "Trump's Defense Secretary Cites Climate Change as a National Security Challenge," ProPublica, March 14, 2017, https:// www.propublica.org/article/trumps-defense-secretary-cites-climate -change-national-security-challenge.

4. Thomas Robert Malthus, *An Essay on the Principle of Population* (1798).

5. Charles Dickens, *A Christmas Carol* (London: Chapman & Hall, 1843).

6. Miriam King and Steven Ruggles, "American Immigration, Fertility, and Race Suicide at the Turn of the Century," *Journal of Interdisciplinary History* 20, no. 3 (1990): 347–369.

7. "A Brief Chronology of the Sierra Club's Retreat from the Immigration-Population Connection," Center for Immigration Studies, August 14,

2018, https://cis.org/Immigration-Studies/Brief-Chronology-Sierra-Clubs-Retreat-Immigration-Population-Connection-Updated; Michael Brune, "Pulling Down Our Monuments," Sierra Club, July 22, 2020, https://www.sierraclub.org/michael-brune/2020/07/john-muir-early-history-sierra-club.

8. Paul Ehrlich, *The Population Bomb* (New York: Ballantine Books, 1968).

9. *The Tonight Show Starring Johnny Carson*, guest Paul Ehrlich, aired January 31, 1980, on NBC, https://www.youtube.com/watch?v=6E5lUNBk3zQ.

10. Alex De Waal, "The End of Famine? Prospects for the Elimination of Mass Starvation by Political Action," *Political Geography* 62 (2018): 184–195.

11. Paul R. Ehrlich and Anne H. Ehrlich, "The Population Bomb Revisited," *Electronic Journal of Sustainable Development* 1, no. 3 (2009): 63–71.

12. "Delhi, India Metro Area Population 1950–2024," Macrotrends, https://www.macrotrends.net/cities/21228/delhi/population.

13. "Global Issues: Population," United Nations, https://www.un.org/en/global-issues/population.

14. Donella H. Meadows, Dennis L. Meadows, Jørgen Randers, and William W. Behrens III, *The Limits to Growth* (New York: Universe Books, 1972).

15. Isaac Asimov, *Second Foundation* (New York: Gnome Press, 1953).

16. Prajakta R. Gupte, "India: 'The Emergency' and the Politics of Mass Sterilization," *Education About Asia* 22, no. 3 (2017): 40–44.

17. Mara Hvistendahl, *Unnatural Selection: Choosing Boys over Girls, and the Consequences of a World Full of Men* (New York: Public Affairs, 2011).

18. Chelsea Follett, "Neo-Malthusianism and Coercive Population Control in China and India: Overpopulation Concerns Often Result in Coercion," Cato Institute, July 21, 2020, https://www.cato.org/policy-analysis/neo-malthusianism-coercive-population-control-china-india-overpopulation-concerns.

19. Thomas Robertson, *The Malthusian Moment: Global Population Growth and the Birth of American Environmentalism* (New Brunswick, NJ: Rutgers University Press, 2012).

20. Margaret Thatcher, "Speech to the Royal Society (Climate Change)," Fishmongers' Hall, City of London, September 27, 1988, transcript, https://www.margaretthatcher.org/document/107346.

21. @elonmusk, "Population collapse is the biggest threat to civilization," 4:13 PM, May 24, 2022, x.com/elonmusk/status/1529193812949614594.

22. Sophie Alexander, Dana Hull, and Graham Starr, "Musk Is Funding Fertility, Population Research in Texas Project," Bloomberg News, August 14, 2023.

23. Population Connection, *Population Connection 2021 Consolidated Financial Statements* (Washington, DC: Population Connection, 2022), https://populationconnection.org/wp-content/uploads/2022/06/Popconn -2021-Consolidated-FS.pdf.

24. Reece Jones, *White Borders: The History of Race and Immigration in the United States from Chinese Exclusion to the Border Wall* (Boston: Beacon Press, 2021).

25. Jones, *White Borders*.

26. John Tanton, "International Migration as an Obstacle to Achieving World Stability," John Tanton, https://www.johntanton.org/articles /mitchell_essay_immigration.html.

27. "John Tanton's Private Papers Expose More than 20 Years of Hate," Southern Poverty Law Center, November 30, 2008, https://www.splcenter .org/fighting-hate/intelligence-report/2008/john-tanton%E2%80%99s -private-papers-expose-more-20-years-hate.

28. Abrahm Lustgarten, "The Ghosts of John Tanton," ProPublica, October 19, 2024, https://www.propublica.org/article/john-tanton-far -right-extremism-environmentalism-climate-change.

29. "WITAN Memo III," Southern Poverty Law Center, January 29, 2010, https://www.splcenter.org/fighting-hate/intelligence-report/2015 /witan-memo-iii.

30. "Federation for American Immigration Reform," Southern Poverty Law Center, https://www.splcenter.org/fighting-hate/extremist-files/group /federation-american-immigration-reform.

31. "NumbersUSA History," NumbersUSA, https://www.numbersusa .com/about/history-of-numbersusa.

32. Jones, *White Borders*.

33. "John Tanton," Southern Poverty Law Center, https://www.splcenter .org/resources/extremist-files/john-tanton.

34. Jones, *White Borders*.

35. Nicholas Kulish and Mike McIntire, "In Her Own Words: The Woman Who Bankrolled the Anti-Immigration Movement," *New York Times*, August 14, 2019.

36. Jean Guerrero, *Hatemonger: Stephen Miller, Donald Trump, and the White Nationalist Agenda* (New York: HarperCollins, 2020).

37. Nils Gilman, "The Coming Avocado Politics," Breakthrough Institute, February 7, 2020, https://thebreakthrough.org/journal/no -12-winter-2020/avocado-politics.

38. Lustgarten, "Ghosts of John Tanton."

39. Imogen Richards, Callum Jones, and Gearóid Brinn, "Eco-Fascism Online: Conceptualizing Far-Right Actors' Response to Climate Change on Stormfront," *Studies in Conflict & Terrorism* (2022): 1–27.

40. Kate Aronoff, "The European Far Right's Environmental Turn," *Dissent*, May 31, 2019, https://www.dissentmagazine.org/online_articles /the-european-far-rights-environmental-turn.

41. Aude Mazoue, "Le Pen's National Rally Goes Green in Bid for European Election Votes," France24, April 20, 2019, https://www .france24.com/en/20190420-le-pen-national-rally-front-environment -european-elections-france.

42. Yang-Yang Zhou and Andrew Shaver, "Reexamining the Effect of Refugees on Civil Conflict: A Global Subnational Analysis," *American Political Science Review* 115, no. 4 (2021): 1175–1196.

43. Todd Miller, Nick Buxton, and Mark Akkerman, *Global Climate Wall: How the World's Wealthiest Nations Prioritise Borders over Climate Action* (Amsterdam: Transnational Institute, 2021), https://www .tni.org/files/publication-downloads/global-climate-wall-report-tni-web -resolution.pdf.

44. Mark Akkerman, "Global Spending on Immigration Enforcement Is Higher than Ever and Rising," *Migration Information Source*, May 31, 2023, https://www.migrationpolicy.org/article/immigration-enforcement -spending-rising.

45. Reece Jones, "Borders and Walls: Do Barriers Deter Unauthorized Migration?" *Migration Information Source*, October 5, 2016, https:// www.migrationpolicy.org/article/borders-and-walls-do-barriers-deter -unauthorized-migration.

9. Vacations for the End Times

1. "2022 Visitor Spending by Prosperity Zone," Visit NC, 2022, https:// partners.visitnc.com/contents/sdownload/73517/file/2022+Visitor+Spend ing+by+Prosperity+Zone.pdf.

2. Christopher Flavelle, "Tiny Town, Big Decision: What Are We Willing to Pay to Fight the Rising Sea?" *New York Times*, March 14, 2021.

3. "Sea Level Rise Viewer," National Oceanic and Atmospheric Administration, https://coast.noaa.gov/slr/#/layer/sce/7/-8400974.90411118 2/4245312.611130581/15/satellite/19/0.8/2050/high/midAccretion.

4. Flavelle, "Tiny Town, Big Decision."

5. European Travel Commission, "Summer 2024: More Europeans Plan to Travel, but Taking Fewer Trips," press release, July 2, 2024, https://etc-corporate.org/uploads/2024/07/ETC_Press-release_Intra-European -Travel-Trends_02-07-final.pdf.

6. "Travel & Tourism Economic Impact Research (EIR)," World Travel and Tourism Council, https://wttc.org/research/economic-impact.

7. Julian Hattem, host, *Changing Climate, Changing Migration*, podcast, " 'Coolcations' and 'Last-Chance Tourism': How Climate Change Is Upending Vacation Planning," Migration Policy Institute, July 23, 2024, https://mpichangingclimatechangingmigration.podbean.com/e /climate-change-vacation-planning.

8. Brittany Van Voorhees, "Beach Erosion the Cause of Home Collapse on the Outer Banks," WCNC Charlotte, February 10, 2022, https://www .wcnc.com/article/weather/beach-erosion-cause-home-collapse-outer -banks/275-3ef646f9-95d4-4f64-903b-3f9513bc36c3.

9. Jeremy Markovich, "The Story Behind an Outer Banks House That Collapsed into the Ocean," North Carolina Rabbit Hole, May 12, 2022, https://www.ncrabbithole.com/p/rodanthe-nc-outer-banks-ocean-house -collapse; Joy Crist, "House Collapses Along Outer Banks Coast," *Island Free Press*, May 10, 2022, https://islandfreepress.org/outer-banks-news /house-collapses-overnight-in-rodanthe.

10. Associated Press, "WATCH: Video Shows One of Two Outer Banks Beach Houses Collapsing into Atlantic Surf," WUNC, May 10, 2022, https://www.wunc.org/news/2022-05-10/another-outer-banks-beach -house-falls-along-the-coast.

11. Brady Dennis, "These Houses Are at Risk of Falling into the Sea. The U.S. Government Bought Them," *Washington Post*, October 16, 2023.

12. John Kowlok, "Case Study 8: Relocating the Lighthouse, Cape Hatteras National Seashore, North Carolina," in *Coastal Adaptation Strategies: Case Studies*, ed. Courtney A. Schupp, Rebecca L. Beavers, and Maria A. Caffrey (Fort Collins, CO: National Park Service, 2015).

13. Brady Dennis, "North Carolina Beach Houses Have Fallen into the Ocean. Is There a Fix?" *Washington Post*, May 15, 2023.

14. European Commission, "Global Warming to Reshuffle Europe's Tourism Demand, Particularly in Coastal Areas," Joint Research Centre News, July 28, 2023, https://joint-research-centre.ec.europa.eu/jrc -news-and-updates/global-warming-reshuffle-europes-tourism-demand -particularly-coastal-areas-2023-07-28_en; Jamey Keaten, "Europe Is the Fastest-Warming Continent, at Nearly Twice the Average Rate, Report Says," Associated Press, April 22, 2024, https://apnews.com/article /copernicus-heat-climate-europe-world-meteorological-organization -d08b3bd028bc461f281f39828bd73056.

15. Lynn Brown, "Will Extreme Weather Change When (and Where) You Go on Holiday?" BBC News, August 1, 2024.

16. Elena Becatoros and Srdjan Nedeljkovic, "Tourist Attractions Close as Extreme Heat Forces Many in Europe to Stay Inside," *The Independent*, July 18, 2024, https://www.the-independent.com/travel/news-and-advice /greece-italy-heatwave-temperatures-athens-b2581890.html.

17. Matthew Pearce, "Michael Mosley's Cause of Death 'Unascertainable,' Coroner Says," *The Guardian*, December 20, 2024.

18. "A Massive Wildfire in Northeastern Greece Is Gradually Abating, with over 700 Firefighters Deployed," Associated Press, September 4, 2023, https://apnews.com/article/greece-wildfires-fd0c877 fc9a2332caf5f1934406af53f.

19. "Flames Devour Forests and Homes as Wildfires in Greece Burn Out of Control," Al Jazeera, August 24, 2023.

20. Elena Becatoros, "Severe Flooding in Greece Leaves at Least 6 Dead and 6 Missing, Villages Cut Off," Associated Press, September 7, 2023.

21. Clea Skopeliti, "A 'Biblical Catastrophe': Death Toll Rises to Six as Storm Daniel Lashes Greece," *The Guardian*, September 7, 2023; Helena Smith, "Greek PM under Attack over Handling of Storm Daniel Disaster Responses," *The Guardian*, September 11, 2023.

22. European Travel Commission, "Summer 2024."

23. Joe Goodman, "Recent Alps Snow Cover Decline 'Unprecedented' in Past 600 Years," Carbon Brief, February 15, 2023, https://www .carbonbrief.org/recent-alps-snow-cover-decline-unprecedented-in-past -600-years.

24. Doyle Rice, "Climate Change Terrifies the Ski Industry. Here's What Could Happen in a Warming World," *USA Today*, January 19, 2024.

25. "Climate Change—Advocacy & Action," National Ski Areas Association, https://nsaa.org/climate.

26. Daniel Scott and Robert Steiger, "How Climate Change Is Damaging the US Ski Industry," *Current Issues in Tourism* 27, no. 22 (2024): 3891–3907.

27. "Climate Change Spurs Innovation at Ski Resorts Adapting to a Warmer World," *Globe and Mail*, January 24, 2023, https://www.theglobeandmail.com/business/industry-news/property-report/article-climate-change-spurs-innovation-at-ski-resorts-adapting-to-a-warmer.

28. Noah Haggerty, "Inside the Battle to Save Mountain High Ski Resort from a Monster California Wildfire," *Los Angeles Times*, October 20, 2024.

29. World Travel and Tourism Council (WTTC), "Greek Travel & Tourism Sector to Approach Full Recovery This Year, Says WTTC," press release, June 22, 2023, https://wttc.org/news-article/greece-eir-2023.

30. Organisation for Economic Co-operation and Development (OECD), *Caribbean Development Dynamics 2025* (Paris: OECD Publishing, 2024), https://www.oecd.org/content/dam/oecd/en/publications/support-materials/2024/12/caribbean-development-dynamics-2024_86de96a9/Overview_Caribbean%20Development%20Dynamics%202025.pdf.

31. Faris Hadad-Zervos, "Beyond Tourism: The Evolving Narrative of Maldives' Growth," *World Bank Blogs*, March 31, 2022, https://blogs.worldbank.org/en/endpovertyinsouthasia/beyond-tourism-evolving-narrative-maldives-growth.

32. N. Biddle, B. Edwards, D. Herz, and T. Makkai, *Exposure and Impact on Attitudes of the 2019–20 Australian Bushfires* (Canberra: Australian National University, Centre for Social Research and Methods, 2020), https://csrm.cass.anu.edu.au/research/publications/exposure-and-impact-attitudes-2019-20-australian-bushfires-0.

33. "Smoke from the Black Summer Wildfires in Australia Impacted the Climate and High Altitude Winds of the Southern Hemisphere for More than a Year and a Half," *ScienceDaily*, September 6, 2022, https://www.sciencedaily.com/releases/2022/09/220906114216.htm.

34. Natasha Schapova, "Black Summer Bushfires Cost Australia Billions in Lost Tourism, New Research Reveals," Australian Broadcasting Corporation, February 12, 2024, https://www.abc.net.au/news/2024-02-13/black-summer-bushfires-economic-impacts-research-snowy-valleys/103446338.

35. WTTC, *Caribbean Resilience and Recovery: Minimising the Impact of the 2017 Hurricane Season on the Caribbean's Tourism Sector* (London:

WTTC, 2018), https://wttc.org/Portals/0/Documents/Reports/2018/Caribbean%20Recovery%20Report%20-%20Full%20Report%20-%20Apr%202018.pdf?ver=2021-02-25-182520-540.

36. Eduardo Cavallo, Santiago Gómez, Ilan Noy, and Eric Strobl, *Climate Change, Hurricanes, and Sovereign Debt in the Caribbean Basin* (Washington, DC: Inter-American Development Bank, 2024), https://publications.iadb.org/en/publications/english/viewer/Climate-Change-Hurricanes-and-Sovereign-Debt-in-the-Caribbean-Basin.pdf.

37. Brown, "Will Extreme Weather Change When (and Where) You Go on Holiday?"

38. Virtuoso, "Virtuoso, the Voice of Luxury, Reveals the Trends and Insights Defining Luxury Travel for the Fall and Festive Seasons and Beyond," press release, August 14, 2024, https://static.virtuoso.com/division-marketing/PR/VTW-2024-Releases/VTW%202024%20Trends%20Release_FINAL.pdf.

39. Condé Nast Traveler and Sarah Allard, "The Biggest Travel Trends to Expect in 2024," *Condé Nast Traveler*, December 30, 2023, https://www.cntraveler.com/story/travel-trends-2024.

40. European Commission, "Global Warming to Reshuffle Europe's Tourism Demand."

41. "Enjoy a Coolcation: Refreshing Travel Ideas for Norway," Visit Norway, https://www.visitnorway.com/plan-your-trip/coolcation.

42. Mark Nicholls, *Climate Change: Implications for Tourism* (Cambridge: University of Cambridge, Institute for Sustainability Leadership, 2014), https://www.cisl.cam.ac.uk/system/files/documents/ipcc-ar5-implications-for-tourism-briefing-prin.pdf.

43. Jackie Dawson, Emma J. Stewart, Harvey Lemelin, and Daniel Scott, "The Carbon Cost of Polar Bear Viewing Tourism in Churchill, Canada," *Journal of Sustainable Tourism* 18, no. 3 (2010): 319–336.

44. John Bartlett, "Antarctic Tourism Is Booming—but Can the Continent Cope?" *The Guardian*, June 25, 2023.

45. Annah E. Piggott-McKellar and Karen E. McNamara, "Last Chance Tourism and the Great Barrier Reef," *Journal of Sustainable Tourism* 25, no. 3 (2017): 397–415.

46. Dandan Liu, Jing Ji, and Meng Wu, "Tourism Carbon Emissions: A Systematic Review of Research Based on Bibliometric Methods," *Journal of Quality Assurance in Hospitality & Tourism* (2023): 1–21; M. Crippa, D. Guizzardi, E. Schaaf, F. Monforti-Ferrario, R. Quadrelli, A. Risquez Martin, et al., *GHG Emissions of All World Countries*

(Luxembourg: Publications Office of the European Union, 2024), https://doi.org/10.2760/4002897.

47. Nicholls, *Climate Change: Implications for Tourism.*

48. Jamie D'Souza, Jackie Dawson, and Mark Groulx, "Last Chance Tourism: A Decade Review of a Case Study on Churchill, Manitoba's Polar Bear Viewing Industry," *Journal of Sustainable Tourism* 31, no. 1 (2023): 14–31.

49. Matthew Harris, "Each Antarctic Tourist Effectively Melts 83 Tonnes of Snow—New Research," The Conversation, February 22, 2022, https://theconversation.com/each-antarctic-tourist-effectively-melts-83-tonnes-of-snow-new-research-177597.

50. Miriam Quick, "Flygskam," BBC, July 22, 2019.

51. Expedia Group Media Solutions, "Travelers' Interest in Sustainable Tourism Options Increases," *Expedia Group Blog*, April 18, 2023, https://advertising.expedia.com/blog/sustainability/sustainable-tourism-demand-rises.

52. Emmanuel Salim, Ludovic Ravanel, and Philip Deline, "Does Witnessing the Effects of Climate Change on Glacial Landscapes Increase Pro-Environmental Behaviour Intentions? An Empirical Study of a Last-Chance Destination," *Current Issues in Tourism* 26, no. 6 (2023): 922–940.

53. Paige McClanahan, "It Just Got Easier to Visit a Vanishing Glacier. Is That a Good Thing?" *New York Times*, March 3, 2024.

10. A New Human Map

1. Michael B. Gerrard, "America Is the Worst Polluter in the History of the World. We Should Let Climate Change Refugees Resettle Here," *Washington Post*, June 25, 2015.

2. "Buffalo Niagara as a Climate Change Refuge," Be in Buffalo, https://beinbuffalo.com/community/climate-refuge.

3. Jeremy Deaton, "Will Buffalo Become a Climate Change Haven?" Bloomberg News, December 5, 2019, https://www.bloomberg.com/news/articles/2019-12-05/the-consequences-of-being-a-climate-refuge-city.

4. Zillow Group, "Buffalo Thunders Back as Zillow's Hottest Market for 2025," press release, January 7, 2025, https://investors.zillowgroup.com/investors/news-and-events/news/news-details/2025/Buffalo-thunders-back-as-Zillows-hottest-market-for-2025/default.aspx.

5. Steven Hubbard, "Immigrant Entrepreneurs and the Fortune 500: Powering the U.S. Economy in 2024," Immigration Impact, Septem-

ber 13, 2024, https://immigrationimpact.com/2024/09/13/immigrant
-entrepreneurs-fortune-500-powering-the-u-s-economy-in-2024.

6. UNHCR, *Mid-Year Trends 2024* (Geneva: UNHCR, 2024), https://
www.unhcr.org/mid-year-trends-report-2024.

7. "The Henley Passport Index: 2025 Global Rankings," Henley &
Partners, https://www.henleyglobal.com/passport-index/ranking.

8. "Golden Visas," Henley & Partners, https://www.henleyglobal.com
/residence-investment/golden-visa.

9. Alex Luke, Ted Christie-Miller, Jack Richardson, and Phoebe Bunt,
Forced to Move: How to Reduce Climate Migration to the UK (London:
Onward, 2023), https://www.ukonward.com/reports/forced-to-move
-climate-migration.

10. Ash Gillis, Nathaniel Geiger, Kaitlin Raimi, Julia Lee Cunningham,
and Melanie A. Sarge, "Climate Change–Induced Immigration to the
United States Has Mixed Influences on Public Support for Climate Change
and Migrants," *Climatic Change* 176, no. 5 (2023): 48.

11. Alexander Betts, *Survival Migration: Failed Governance and the
Crisis of Displacement* (Ithaca, NY: Cornell University Press, 2013); John
Washington, *The Dispossessed: A Story of Asylum and the US-Mexican
Border and Beyond* (New York: Verso, 2020).

12. James Kanter and Andrew C. Revkin, "World Scientists Near Consensus on Warming," *New York Times*, January 30, 2007.

13. "The Funds," Climate Funds Update, updated January 2025, https://
climatefundsupdate.org/the-funds.

14. United Nations Environment Programme, *Adaptation Gap Report
2024* (Nairobi: UNEP, 2024), https://www.unep.org/resources/adaptation
-gap-report-2024.

15. Julian Hattem, host, *Changing Climate, Changing Migration*,
podcast, "The Reverse of Climate Migration: Should There Be a Right
Not to Be Displaced amid Climate Change?" Migration Policy Institute,
November 15, 2024, https://www.migrationpolicy.org/multimedia
/reverse-climate-migration-should-there-be-right-not-be-displaced-amid
-climate-change.

16. Anil Markandya and Mikel González-Eguino, "Integrated Assessment for Identifying Climate Finance Needs for Loss and Damage: A Critical Review," *Loss and Damage from Climate Change* (2019): 343–362;
United Nations Climate Change, "Pledges to the Fund for Responding to
Loss and Damage," UNFCCC, updated April 7, 2025, https://unfccc.int

/topics/climate-finance/funds-entities-bodies/fund-for-responding-to-loss
-and-damage/pledges-to-the-fund-for-responding-to-loss-and-damage.

17. Lily Katz and Sebastian Sandoval-Olascoaga, "More People Are
Moving In Than Out of Areas Facing High Risk from Climate Change,"
Redfin News, August 25, 2021, https://www.redfin.com/news/climate
-migration-real-estate-2021.

18. Lily Katz and Sheharyar Bokhari, "Migration Into America's Most
Flood-Prone Areas Has More Than Doubled Since the Start of the
Pandemic," Redfin News, July 24, 2023, https://www.redfin.com/news
/climate-migration-real-estate-2023.

Index

About the Author

Julian Hattem has been a journalist, writer, and editor focused on politics, government, and migration for more than fifteen years. He has been on staff with the Associated Press, *The Hill*, and the *Yomiuri Shimbun*, and has written for outlets including the *Washington Post*, *The Guardian*, NPR, and *The Atlantic*. He has reported from four continents and is currently the editor of the *Migration Information Source*, the online magazine of the Migration Policy Institute, and founder and host of the podcast *Changing Climate, Changing Migration*. *Shelter from the Storm* is his first book. He lives in Washington, DC.

Publishing in the Public Interest

Thank you for reading this book published by The New Press; we hope you enjoyed it. New Press books and authors play a crucial role in sparking conversations about the key political and social issues of our day.

We hope that you will stay in touch with us. Here are a few ways to keep up to date with our books, events, and the issues we cover:

- Sign up at www.thenewpress.com/subscribe to receive updates on New Press authors and issues and to be notified about local events
- www.facebook.com/newpressbooks
- www.x.com/thenewpress
- www.instagram.com/thenewpress

Please consider buying New Press books not only for yourself, but also for friends and family and to donate to schools, libraries, community centers, prison libraries, and other organizations involved with the issues our authors write about.

The New Press is a 501(c)(3) nonprofit organization; if you wish to support our work with a tax-deductible gift please visit www.thenewpress.com/donate or use the QR code below.